The Remotest Island

Solving the riddle of the flightless
moorhen of Tristan da Cunha

Albert J. Beintema

Published by New Generation Publishing in 2022

First Edition

Paperback ISBN: 978-1-80369-254-8
Hardback ISBN: 978-1-80369-255-5

www.newgeneration-publishing.com

New Generation Publishing

The Remotest Island

Solving the riddle of the flightless
moorhen of Tristan da Cunha

edited by Sue McVerry

*Dedicated to Dineke (1946-2003), who finally joined me
on my third voyage to Tristan da Cunha*

Sailing on the winds of adventure,
Across a sea of loneliness,
To a destination and a dream,
that awaits there for you.

Found pinned on the notice-board in the
school at Tristan da Cunha (1993)

Contents

Introduction

The Remotest Island, is a modified remake of my earlier book, *Het Waterhoentje van Tristan da Cunha* (The Moorhen, or Gallinule, of Tristan da Cunha), which was published in Dutch, in 1997. Although the Moorhen is the recurrent theme in this book, most of the text was dedicated to the island's geography, its history, and its people. Much has been written on Tristan in English, so in the present version, I write about history much more briefly, concentrating on the juiciest stories, except where I have to add new information.

When I published my book in 1997 I had only been to Tristan once. Since then, I have been back many times, quite unexpectedly, so I have a lot of new stories to tell, even about the wretched Moorhen.

I have to make a special note about the Tristanians and their sensitivities. Tristan people don't usually like others writing about them, or making films about island life. They argue that during a brief visit outsiders cannot possibly form a true picture. In their view, such outsiders are far too quick to ridicule island life, and portray islanders as an ignorant, inbred, sorry lot on their remote piece of rock in the middle of the ocean. In reality, of course it is these outsiders who are the ignorant ones. Tristan people like to make fun of each other, themselves, and outsiders, but they are extremely sensitive when outsiders make fun of them. This is understandable, as they have had some difficult experiences with publicity, especially during their involuntary stay in England, following their evacuation due to the volcanic eruption in the sixties. These experiences were so unpleasant that they could not wait to return to their lonely isle.

One would think that living in such a remote place would guarantee freedom from inquisitive people elsewhere. The truth is different. Tristan people live in a figurative terrarium, being constantly closely observed by those curious outsiders. Hundreds of people, all over the world, have an inexplicable interest in the island, and keep a constant close watch. They want to know how Tristanians live, how they speak, what they eat, with whom they sleep, and so on. This helps to explain the wariness that Tristan people display towards outsiders. There are several films made about the island, and many books have been written. Tristanians make a strict division between the good movies and the bad movies, the good books, and the bad ones. Most are bad in their view. One writer who chose the wrong topics was subsequently refused access on a later visit. The only ones passing the test are those which picture the island community as a sugar sweet group of extremely nice, friendly people. Departing visitors are supposed to leave behind three hundred best friends.

My experiences are mixed. I found the people of Tristan da Cunha extremely ordinary, just like in any small village in coastal Scotland or Ireland. I did not leave behind three hundred best friends. I just found some people nicer than others, as I would in any other village, anywhere else in the world. Inevitably Tristan people have a few peculiarities, which lend themselves to an amusing story but that does not mean I discriminate or pass judgement. I would do exactly the same in any other community I might write about, including my home town. I hope the people of Tristan will appreciate this.

In fact, I make more fun of various outsiders, who want so badly to meddle with Tristan affairs, including myself, being a nosy outsider *pur sang*.

This is not a scientific book, so in most of the text, I refrain from giving references to sources, other than listing them at the end. I make an exception for the chapters on shipping in the old days, and the history of the Moorhen, where I feel references are a necessity, and I only occasionally add one elsewhere.

Part 1. Tristan da Cunha, a Dream

1. A Dream

I heard Tristan da Cunha referred to as a 'destination of the mind' which rang very true. The tiny island of Tristan da Cunha, the most remote inhabited destination on earth, 2800 km west of cape Town, 4000 km east of Buenos Aires, right in the middle of the vast, empty expanse of the South Atlantic. Tristan da Cunha, the island with the most beautiful name in the world.

I know many people who are infected by what is known as the Tristan Bug, often for the silliest reasons. Indeed I am a severe case myself. I think I caught the bug when I was given my first true bird guide for my 15th birthday: Peterson's excellent *Field Guide to the Birds of Britain and Europe*. Until then I used some mediocre field guides I found on my father's book-shelves. I was very proud when I managed to identify bird species my father had never seen. Roger Tory Peterson changed the world for the amateur birdwatcher.

I became an avid reader of Peterson's Field Guide, finding exotic and colourful birds like Bee-eaters and Hoopoes. I dreamt of seeing them in the wild, in equally exotic countries. And then I found the most intriguing bird of all: the Greater Shearwater. This was a true oceanic bird, only coming to land in order to breed, and at that time only known to nest in Tristan da Cunha. Oh dear! There was an estimated 2,000,000 pairs and just about 2,000,000 of those were packed together on the tiny uninhabited island of Nightingale - another island with a beautiful name. A few hundred thousand would nest in other uninhabited islands in the Tristan Group, like Inaccessible and Gough. Nightingale measures only one by one-and-a-half kilometre, and hosts vast numbers of other seabird species as well. A birder's paradise!

After breeding, the shearwaters leave Tristan and fly east towards South Africa, using the prevailing westerlies at their latitude. Before seeing land the birds turn to the left to pick up the Southeast Trade Winds which help them to cross the South Atlantic diagonally all the way to the tropical waters off Brazil. Then they must struggle through doldrums and the Northeast Trade Winds in the northern Tropics until they pick up the southwesterlies in the North Atlantic. These help them to reach the cold, nutrient rich waters around Greenland and Iceland, where they spend their southern winter during our northern summer.

On the way back to Tristan they pass the waters around the British Isles, where fanatical bird watchers lie in wait to see them and then, off the coast of West Africa, they return to the Brazilian seas, once more using the Trade Winds. Finally, following the coasts of Brazil and Argentina, they find the southern westerlies, which bring them back to their tiny island home, thus completing their gigantic figure-eight migration route. It is a true miracle that these shearwaters are able to find that tiny speck in the middle of the ocean! As if they have a compass and a GPS in their little brains. Well, they actually do have something like that, but instead of satellites they cleverly use the earth's magnetic field. I found the Greater Shearwater the most interesting and mysterious bird I ever heard of. Well worth a dream. I think is was then and there that I decided that my future would lie in studying oceanic birds in general, and those of Tristan da Cunha in particular. Tristan became my prime dream destination.

As a true vague dreamer I never took the trouble to read anything about Tristan da Cunha. So during the early sixties, dreaming away, I totally missed the headlines about the volcanic eruption on Tristan and the dramatic evacuation that followed. My dream remained a hollow one. Until, in the late sixties, I met the girl who would become my wife, Dineke. She suggested we should visit the library to find books about Tristan. And so we did. She became equally intrigued and soon we decided to go to Tristan together. We wrote a fantastic plan for a combined PhD thesis in which she, as a botanist, would map the vegetation of Nightingale, and I, as an animal ecologist, would study the seabirds. Together we would analyse the interaction between plants and birds. Beautiful! We would go to Tristan for two years, spending one whole year in isolation on tiny Nightingale Island. I designed special wooden packing crates for all our equipment and food rations, which when unpacked and disassembled, could be rearranged to build the hut we would live in during our stay on Nightingale.

Our Great Expedition never came to fruition. The plan was more romantic than scientifically sound, and we (our Professor and the two of us) failed to raise the necessary funds. In the end we had to give up and do other things to stay alive. The Great Dream remained just that, and over the years it slowly faded away. I wrote an alternative PhD thesis on farmland birds such as Lapwings and Black-tailed Godwits, we had children, and led busy lives. Tristan was further away than ever. But the Dream never completely died. Deep inside, I knew that one day I would set foot on Tristan.

2. Just a Moorhen

When I finally managed to get to Tristan it was not to study the seabirds of Nightingale. I found another excuse. I went to Tristan to solve the mystery of the flightless Tristan da Cunha Moorhen, known locally as Island Cock or sometimes Island Hen.

Superficially there is nothing special about the Tristan Moorhen. It looks just like any other ordinary Moorhen, that we might see in our parks and ponds. Blackish brown, with a row of white specks along the flanks, a yellow and red bill, yellow or greenish legs, and a nervously twitching cocked tail with a few white feathers on the underside. Perhaps it is a little darker than our local Moorhen at home, the legs are stronger and thicker, and the wings are shorter and more rounded. But these small differences do not make it special. The thing that makes it special is that it is, or was, to be found nowhere except on Tristan da Cunha. Whether we say 'is' or 'was' depends on whether we consider the endemic Island Cock extinct or not. There was even a question as to whether it ever existed at all. This was the Great Mystery I was going to solve.

The original Moorhen of Tristan was unable to fly. The wings looked normal but the primaries were too soft, the flight muscles too weak to lift the bird off the ground. Loss of flight ability is a feature often seen in birds on isolated islands The most famous example is the extinct Dodo of Mauritius. In the rail family, to which Coots and Moorhens belong, flightlessness on islands is particularly common. Rails in general are poor flyers. Perhaps this is why they are prone to being blown away from their migration routes when strong winds come from the wrong direction. When they are blown away from the land these lost birds will usually perish at sea unless they accidentally find an island. It is easy to imagine that an involuntary, exhausting flight over thousands of miles of open ocean, would be an extremely traumatic experience for a land bird and that, once safely on an island, they might make a vow never to fly again. However, bird brains do not work like that but there are certainly evolutionary advantages to becoming flightless in the absence of predators. Today, most flightless island species are either already extinct or threatened with extinction, due to the arrival of Man and the accompanying goats, pigs, cats and rats. Allegedly the Tristan Moorhen went the way of the Dodo in the nineteenth Century.

The Tristan da Cunha Moorhen was described as a new species to science in 1861. Sir George Grey, Governor of the Cape Colony, donated a fine collection of live animals to the Zoo of the Zoological Society in London. The animals were to be shipped from Cape Town to London under the care of Mr Benstead, the Society's agent. A few days earlier, a Tristanian girl working in Sir George's household, brought five live Moorhens from her island to Cape Town. They joined the South African collection in the ship's hold. Unfortunately, their cage was placed a little too close to another cage which held a hungry Jackal. Already before departure the Jackal had killed two Moorhens. Mr Benstead prepared the skins, but, alas, the heads were missing having been eaten by the Jackal. Two other Moorhens died during the long sea voyage. They were put in jars and preserved with alcohol. Eventually number five arrived alive and well in England, where it lived happily for three years in the London Zoo. In 1864 it died and ended up between the camphor balls in a drawer of the British Museum.

In 1861 Doctor Philip Lutley Sclater, Secretary of the Zoological Society, examined the four dead Moorhens and the live one in the Zoo. He concluded these were not ordinary Moorhens, but a new species, as yet undescribed. A scientist describing a new species has to follow strict international rules, based on a system designed by Linnaeus in the mid-eighteenth century. Every species gets two Latin names, the first indicating the Genus, the second the Species. A Genus contains one or more closely related Species. Our Common Moorhen is called *Gallinula chloropus* - greenlegged little chicken. Sclater found that the Tristan Moorhen looked enough like an ordinary Moorhen (or Gallinule), to be placed in the same Genus *Gallinula*. In the Species name he wished to reflect the insular origin of the bird. In true Latin this would have become *insularis* but for reasons unknown Sclater preferred to use the Greek word *nesiotis* (in his description he adds in parentheses 'nesiotis = insularis'). So, in full, the scientific name of the Tristan Moorhen became *Gallinula nesiotis*.

Old reports tell us that Tristan Moorhens tasted delicious. They were not only eaten by Man, but also by the dogs, cats and rats that came to the island with them. Their combined efforts managed to wipe out the species before the end of the nineteenth Century. In 1873 the famous Challenger Expedition called at Tristan da Cunha. The naturalists knew about the Moorhens and wished to see them in the wild. But the islanders told them that Island Cocks were no longer to be found anywhere near the Settlement. The Expedition failed to observe them. In spite of their lack of effort, the negative accounts of the Challenger Expedition caused the Tristan Moorhen to be declared officially extinct by 1873.

All that is left of the Tristan Moorhen is a handfull of bones and two prepared skins - the Type Specimen, and the Co-Type, respectively. The scientific history of the species is an extremely short one: from 1861 until 1873 - just twelve years. An insignificant history of an insignificant brown bird that was doomed to be forgotten.

In 1888 a new Moorhen was discovered, by George Comer, second mate of the American whaler *Francys Alleyn*, when they were killing seals at Gough Island. Like the Tristan Moorhen it looked very much like an ordinary Moorhen, with a cocked tail white underneath, white flecks on the flanks, and yellow legs. Comer called the birds 'Mountain Cocks', and said:

"They cannot fly and use their wings to help them in running... They are quite plentiful and can be caught by hand. Could not get on a table three feet high. The bushes grow on the island up to about 2000 feet, and these birds are found as far up as the bushes grow... Tip of bill bright yellow, scarlet between the eyes. Legs and feet yellow, with reddish spots."

Comer took six birds on board alive, four of which quickly died. Two reached America alive and were kept there for a while, tethered by a rope-yarn, but eventually they escaped. The skins of the four dead ones ended up on the desk of scientist J.A. Allen, who described them as a new species in 1892. He found the weak wings too different from those of the Common Moorhen to place the bird in the same Genus so he designed a new Genus name, *Porphyriornis*. The Species name, he derived from Comer. So the full name of the Gough Moorhen became *Porphyriornis comeri*. Allen also studied the thirty year old skin of Sclater's Type Specimen, finding it so similar to his Gough Moorhen that he had to place the two species in the same Genus. Thus, he rechristened the Tristan Moorhen and called it *Porphyriornis nesiotis*.

Gough Moorhens never became extinct, and are still to be found all over the island. Specimens have been taken regularly, both alive and dead. Live Gough moorhens could be seen in the London Zoo, where they bred happily for many years. In England Gough Moorhens have escaped and are now running wild, interbreeding with the closely related Common Moorhen. The Gough Moorhen is a thriving species currently not threatened with extinction.

In 1970, Gisela Eber, a German biologist, studied the zoogeography of the Moorhens of the world. The Common Moorhen is a cosmopolitan species, to be found in almost every continent, and on many oceanic islands. When comparing populations, carefully measuring skins in museums, one can analyse differences and resemblances and thus form an idea about the history of the species' distribution, and its geographical origin.

Digging her way through the ornithological literature, searching for remote Moorhen populations, Gisela inevitably came accros the Moorhens of Tristan and Gough. She also studied skins of both species in the British Museum, where, at that time, she only could find one single skin of the Tristan Moorhen. She concluded that there was no difference between the two species and that the differences between the descriptions of Sclater and Allen all fell within the individual variation in a longer series of skins from Gough Island. Thus, she combined the two forms as one single species. She also disagreed with Allen that abberant wings justified a separate Genus status. She brought the Tristan/Gough moorhen back to *Gallinula*, where all ordinary Moorhens are to be found.

Finally, Gisela reasoned that is was extremely unlikely that a single species, colonising two seperate islands, and evolving into flightlessness on both, would not diverge through chance processes and under the pressure of markedly different climatic contitions (Gough being much colder and wetter than Tristan). From an evolutionary viewpoint this might even be considered an impossibility. When she found that the Tristan Moorhen only 'existed' from 1861 to 1873, just yielding the one specimen she could find, and that it was followed up by the Gough Moorhen in 1892, her conclusion was logical: The single specimen allegedly from Tristan, must in reality have come from Gough, and in all probability was mislabelled. It was not at all unlikely that people who collected specimens on Gough Island, would label them as 'found near Tristan da Cunha.'

Gisela Eber was not the first scientist to cast doubt on the validity of the Tristan Moorhen as a species that really existed. In 1949 two South African marine biologists visited Tristan to investigate the possibility of starting a viable fishery for Tristan Rock Lobsters (which has been in South African hands ever since). They were also keen ornithologists, and published an extensive paper on the ornithology of Tristan da Cunha. Their opinion on the Moorhens was that is was questionable, at the least, whether the Tristan Moorhen ever existed in reality. But it was Eber who took the firm decision on this issue. She concluded that the Tristan Moorhen was simply a mistake.

The scientific name of the Island Cock has been moved to Gough Island. The Gough moorhen should now be called *Gallinula nesiotis* – Sclater's name, and not *comeri*, because the name *nesiotis* was given earlier. The rules in nomenclature are very strict. With this move the Tristan Moorhen was totally annihilated. It had been reduced to an administrative error.

The Tristan Moorhen was officially 'discovered' as a new species in 1861, but it had been seen by many non-scientific people long before that. Eber, and the scientists who successively believed her, did not know that. But I knew. So I made it my life work, my personal Odyssey, to rehabilitate the Tristan Moorhen, to put it back in history.

3. A brief sketch of the islands

At thirtyseven degrees south and twelve degrees west, Tristan da Cunha lies in the middle of the emptiest ocean in the world, the South Atlantic. As such it is the most remote inhabited place on earth. The nearest human settlement is on the almost equally remote Saint Helena, more than 2400 km to the north, in the tropical South Atlantic. Like Saint Helena, Tristan da Cunha is British territory, and the two islands, in spite of being so far apart, are considered one colony, together with Ascension Island which is almost 1300 km north of Saint Helena. We are not supposed to use the word 'colony' any more. The islands are UKOTs, United Kingdom Overseas Territories. The governor of this one lives on Saint Helena and rarely visits the other islands. These are ruled by an Administrator, and Tristan is supported by an elected Island Council.

The people speak English, but the language has eroded a bit over the generations. Like in the Dutch of South Africa, they lost the conjugations: "I is, you is, we is." If you admire their ability to balance on the boards of small boats, tossed around on the waves, they say "we is used to it". The language has other peculiarities and special words, inherited from first settlers of different nationalities. They do not pronounce the H at the beginning of a word, but in contrast, put one in front a word starting with a vowel. So the islanders are 'highlanders', and my name is 'Halbert', and I live in 'Olland'.

Tristan is inhabited by around 250 people in seven or eight families, derived from only a handful of original settlers. The founder was William Glass, who joined a British Garisson that occupied the island in 1816. When the soldiers went home, Glass stayed behind with his wife. They produced sixteen children. Four bachelor friends obtained wives from Saint Helena, and they too produced a lot of children. Consequently, in the years to come, beautiful young girls enticed shipwrecked or defecting sailers to stay and marry into the population. The people of Tristan lead a peaceful life. The men fish, the women knit, and they all grow potatoes which constitute their staple food. The main source of income is from Rock Lobster fishery. The Rock Lobster is a clawless species of crayfish, or spiny lobster, *Jasus tristani*, only to be found in the waters of Tristan da Cunha and the Vema sea mount which lies around 1600 km northeast of Tristan. This is an ancient volcanic island, eroded away a long time ago, which no longer reaches the surface of the sea, but still carries kelp beds with the associated fauna. A similar species, *Jasus paulensis*, lives in the waters around Saint Paul and Amsterdam Island in the southern Indian Ocean, at about the same latitude as Tristan, and just as far east of Cape Town as Tristan lies west. Recent DNA studies have shown that these two species are in fact identical, and because *paulensis* was described earlier than *tristani*, the Tristan Rock Lobster should be renamed *Jasus paulensis*.

A second source of income comes from the sale of postal stamps, which are issued a few times per year. Tristan stamps are sought after worldwide. Before the fisheries started in the 1950s, the islanders hardly knew money. Business with passing ships was usually done by bartering: shoes, clothes, nails, tobacco, and alcohol against meat, wool, vegetables, potatoes, and souvenirs, like the so-called tasselmats which were made of the scalps of Rockhopper Penguins, with the beautiful yellow plumes radiating. Nowadays, the people can spend their money in the supermarket which is the only shop on the island. Here they can buy goods imported from South Africa, but for a part, they still have a subsistence economy. The islanders grow their own potatoes and a few vegetables, they fish, and they keep cattle, sheep, and poultry.

Tristan can only be reached by ship, a few times a year. Air traffic is not possible. Landings used to be made on the beach below the Settlement. However, the entire bay was covered with lava during the 1961 eruption. Since then, a small harbour has been built which can only accomodate small boats, and is often unusuable when the sea is too rough.

Of the four islands in the archipelago, Tristan, Inaccessible, Nightingale, and Gough, only Tristan, the largest of the four, is inhabited. Tristan is a schoolbook example of a volcano, rising from the seafloor at 3500 m depth, to 2067 m above sea level. A classic, cone-shaped island, almost circular in form, with a diameter of about 11 km. The entire island has been lifted by tectonic movement, causing the sea to erode all sides into sheer, almost vertical cliffs, 300-600 m high. Above that the island slopes gently, gradually becoming steeper towards the summit. This gives the island the shape of an old-fashioned circus tent. At the foot of the cliffs, small plateaus have been formed by secondary eruptions. The largest of these is the Settlement Plateau, on the northwestern side of the island. Here the people live, on a strip of land only six kilometres long and half a kilometre wide. The village, officially known as Edinburgh of the Seven Seas, but usually just called the Settlement, lies on the northern tip of the plateau. There is only one road leading out of the Settlement, to the Potato Patches, a few kilometres to the southwest.

Tristan is the youngest of the four islands at only about 200,000 years old. The volcano was believed to be extinct, until the eruption in 1961 proved otherwise. There was probably an eruption in the southeastern part of the island about 400 years ago, forming a barren area now called Stony Hill. In 1724 a ship reported floating pumice north of the island, indicating another eruption. And in 1852, when the island was already inhabited, a visitor to the south side felt a tremor. Geologists flying over the island in a helicopter in 1984, saw an active steam vent south of the peak. Finally, a submarine eruption was recorded in 2004. So the volcano is certainly not extinct.

The vegetation of Tristan is dominated by the endemic tree fern *Blechnum palmiforme*, which the islanders call Bog Fern, and the shrubby Island Tree *Phylica arborea*, which form dense bushes in de gullies, and on the higher slopes. The Island Tree was thought to be endemic for the Tristan archipelago, but it appears that the *Phylica* trees of Amsterdam Island are the same species, as was the case with the Rock Lobsters. The low plateaus were originally covered in *Spartina arundinacea*, a tall, tough species of Tussock Grass, but in the last two centuries, these vegetations have been destroyed by grazing and burning. This species also occurs both in the Tristan archipelago and the islands of Amsterdam and St. Paul.

Above the cliffs, the gently sloping plain, which the islanders call the Base, a variety of ferns and grasses are found, and in sheltered places, especially on the east side of the island, the trees may grow to a height of several metres. On the western side the trees have mostly been removed for firewood. The tree ferns can grow to a metre and a half in sheltered places, but are usually much smaller, often only knee high. Their trunks can grow in all directions, making walking almost impossible in places. The Island Trees are even worse, often growing horizontally, and branching at ground level hidden by ferns. Together they form the sometimes impenetrable fern bush. At about 1000 m altitude, the fern bush gives way to grasses. Here, the islanders keep sheep herds. As a result of grazing, most grasses found here are introduced species, and some places appear red from Sheep's Sorrel, also an alien species. At about 1500 m, the slopes become steeper and the grasses are replaced by alpine tundra. This part of the island is called the Mountain. Above 1800 m, the Mountain essentially becomes a steep heap of loose cinders and scoria, almost bare, with only small mosses and lichens, culminating in the peak, which holds a crystal clear crater lake. There is snow on the peak most of the year, and in the winter also on the lower slopes.

Tristan has a cool temperate oceanic climate, with an

average temperature at sea level of 15°C, ranging from about 19°C in summer to 12°C in winter. On sunny summer days, the temperature may reach 25°C. In summer the sea, with a temperature of around 18°C, has a cooling effect, while in winter, with a temperature of 13°C, it has a warming effect. Annual rainfall at sea level is in excess of 1500 mm, but on the Mountain may exceed 3000 mm. The Mountain is almost always shrouded by clouds, which with the prevailing, humid winds from the NW, usually already start at 500 m, which is below the edge of the cliffs.

The Mountain and the Base are carved by gullies, running down from the peak radially, like the spokes of a wheel, cutting steep ravines through the cliffs. The islanders call them gulches. In spite of the heavy rainfall, these do not flow constantly as the mountain is very porous. There is only one permanent stream on the island, originating in a spring at the bottom of the cliff, and running accross the Settlement Plateau which provides drinking water for the people. This stream is called the Big Watron. When it is not raining at sea level, showers on the mountain are indicated by waterfalls which suddenly appear over the edge of the cliffs.

On the east side of the island is Sandy Point, the most sheltered, driest, and warmest part of Tristan. Here, the islanders have planted a small forest, with pine trees for timber, apple trees, and even peaches. And there are some cattle grazing, in a semi-wild state. Sandy Point can be reached from the Settlement by crossing the mountain, or, when the sea allows, along the beach at the foot of the cliffs. The easiest way, of course, is by boat, as the landing on the sandy beach is easy, weather permitting. There are also small plateaus in the southeast, near Stony Hill, and the southwest, at The Caves, where cattle graze.

Tristan was once host to millions of seabirds, but most species have been brought close to extinction by man, cats, dogs, and rats. Feral dogs and cats no longer exist. People still keep dogs, but no cats. Rats are everywhere.

Rockhopper Penguins survive in small colonies away from the Settlement, but have been decimated in the past for their eggs and feathers. They are related to the Rockhoppers of the Falkland Islands, but with their much longer yellow head plumes, they have been promoted to a separate species, the Northern Rockhopper. Like the Island Tree and the Rock Lobsters, they are also found on Amsterdam Island and Saint Paul in the Indian Ocean. Tristan used to have three species of albatross: the Yellow-nosed Albatross, locally known as Molly, the all dark Sooty Albatross, known as Peeo, and a large Albatross, formerly thought to be a race of the Wandering Albatross, but now separated as Tristan Albatross. Again, Amsterdam has Mollies, Peeos and great albatrosses too, but the Mollies have now been split into an Indian Ocean species and an Atlantic species, endemic to the Tristan Group. And, as on Tristan, the great albatross of Amsterdam has been promoted to a separate species, the Amsterdam Albatros, now one of the rarest, and most difficult albatrosses in the world to see. The Mollies of Tristan still survive in large numbers, mainly on the eastern side of the island. On the western side they have been decimated, their chicks being considered a delicacy. The Tristan Albatrosses on Tristan have become extinct. The main population is now living on Gough Island, and one or two survive on Inaccessible. The smaller petrel species on Tristan have become very rare, but survive here and there in small numbers.

Two species of tern nest on the island. The Antarctic Tern has its most northerly station in the world here, and the Brown Noddy, a tropical species, has its most southerly breeding place. Tristan is the only place in the world where an Antarctic and a tropical species nest side by side.

Tristan had two songbirds, endemic to the archipelago, the Tristan Bunting and the Tristan Thrush, both also found on Inaccessible and Nightingale, in different races. The Tristan race of the bunting is extinct, as its habitat of tussock fields was destroyed. There is only one skin left in a museum, in Berlin. The thrush is still there, but only in small numbers. And there was (or was there?) the flightless Moorhen, the Island Cock.

Nightingale Island, the smallest of the four, lies 40 km SSW of Tristan, and 25 km south of Inaccessible. It is only one by two km in size, the highest point not reaching 400 m. Therefore the island is too low to catch or develop clouds, and consequently has considerably less rainfall than Tristan. Nightingale is much older than Tristan, an estimated 18 million years. It is the heavily eroded remnant of a much bigger volcano. The island has a friendly, hilly character and is almost completely covered in Tussock Grass. Only in a sheltered valley near the top of the island, *Phylica* woodland is found. Here, the trees grow into real trees reaching six metres. The island is completely perforated with millions of burrows, where all sorts of petrels nest, including the 2,000,000 pairs of Greater Shearwater, known by the islanders simply as 'Petrels'. But there are also about 15 other petrel species breeding, one of the most obvious being the Broad-billed Prion, which is heavily preyed upon by Brown Skuas. The area is littered with their wings. To avoid predation, the Prions, and other small petrel species are strictly nocturnal, but apparently are not always successful at avoiding the Skuas.

Like Tristan, Nightingale has a thrush and buntings. The thrush is the same species as on Tristan and Inaccessible, but a different subspecies, with a more reddish tinge. There are two species of buntings on Nightingale. The story of the buntings is a bit like the Galapagos Finches of Darwin. In the Galapagos Islands, the Finches evolved into different species, all looking alike, but with different bill shapes and sizes, each adapted to a special selection of food items. Due to their different bill sizes they could avoid competition. If on one island, there were no large-billed species, the smaller billed birds would develop slightly larger beaks. In the Tristan archipelago, a similar thing happened. Nightingale has two species, the large-billed Wilkins' Bunting, confined to the *Phylica* woodland, and the small-billed Nightingale Bunting, found all over the island, but mostly in the tussock at lower levels. Tristan had only one, the Tristan Bunting, with an intermediate bill size. With a world population of about 120 pairs, Wilkins' Bunting is one of the rarest birds in the world.

Nightingale has two small satellite islands, Stoltenhoff, and Middle Island (also called Alex Island), both, like Nightingale itself, hosting thousands of Rockhoppers.

Inaccessible Island lies about 40 km southwest of Tristan. It is a square island, with a surface of about 14 square kilometres, and a diameter of about four km. Like Tristan, it is surrounded by almost vertical cliffs, reaching sometimes to more than 500 m. No wonder the island is called Inaccessible. Above that, there is no mountain, as in Tristan, but a fairly level plain, the highest point being Swales Fell, at 600 m. On the east side, there is a small plateau, where safe landings can be made, but from there it is virtually impossible to climb to the interior of the island. The only place where the climb is fairly easy is on the west side, where landing is rarely possible. Like Tristan, Inaccessible usually has cloud cover above 500 m. Being three to four million years old, the island is between Tristan and Nightingale in age.

The vegetation is much like Tristan, with Bog Ferns, Island Trees, and Tussock Grass. The most remarkable plant of Inaccessible is the Tristan Pepper Tree *Peperomia tristanensis*, which in the whole archipelago is confined to one valley on the west side of this island, and which elsewhere is only found in the Chilean Juan Fernandez Islands in the Pacific Ocean. A highly disjunct distribution, with a gap of more than 5000 km!

Inaccessible had feral goats and pigs, but fortunately, they have been exterminated. Best of all, there are no rats.

That leaves the island in a near pristine condition, hosting millions of seabirds. There are even still two or three pairs of the Tristan Albatross. The most remarkable seabird is the Spectacled Petrel (or Spectacled Shoemaker), which was formerly believed to be a subspecies of the White-chinned Petrel (Shoemaker), a prolific species with a circumpolar distribution in the southern oceans. However, on account of its aberrant facial pattern (white spectacles) and a different voice, it has been made a separate species, which is endemic to Inaccessible, not to be found on the other islands.

Inaccessible also has its own subspecies of the same Tristan Thrush, and buntings. The situation with the buntings is a bit more complicated than on Nightingale or Tristan. Like Nightingale, Inaccessible has large-billed and small-billed forms. Originally, the large-billed ones were classified as Wilkins' Buntings, like the ones on Nightingale, but as a different subspecies. The small-billed ones were called Tristan Buntings, also a separate subspecies. The problem is that on Inaccessible the two forms interbreed, which they do not at Nightingale.

There are two models for the evolution of birds on islands. The classic approach, mainly developed in the 1930s, is that geographic isolation is always needed for speciation. If a bird colonises two islands, which can happen if a large flock is blown away from the continent, they will diverge, due to chance mutations, which will be different on the two islands, helped by differences in environmental pressure. Eventually, they will change so much, that they become seperate species. If after that the birds happen to reach the other island, they would no longer recognise each other, not interbreed, and as a result of competition, drift even further apart. This is thought to be how the many species of Galapagos Finches developed. This speciation through isolation is called allopatric speciation. In this model, the buntings at Nightingale have become sufficiently different to behave as two species, but on Inaccessible they have not. Interbreeding will then, ultimately, lead to merging of the two forms, becoming one species again.

The alternative is sympatric speciation, a more recent theory, where in one location a species may split up, if there is variation in size or colour, and the birds show preferential mating, that is birds seeking mates that look like themselves. In that case, size and colour differences will increase, until we end up with different species. In this model, the different forms on Inaccessible are not merging, but, on the contrary, become more differentiated, ultimately leading to seperate species.

Peter Ryan, professor at the Percy Fitzpatrick Institute at the University of Cape Town, spent six months with his wife on Inaccessible (yes, I am jealous), to study the buntings for his PhD. He adhered to the sympatric model, and postulated that the Inaccessible Bunting is still one species, which is in the process of splitting up, not just in two but three forms: the large-billed ones in the *Phylica* wood on the high plateau, a drab small-billed form in the tussock at sea level, and a more brightly coloured one in between. Hybrids between all three still occur, but in the end, Ryan believes, they will become three different species.

And last, but certainly not least, there is the diminutive Inaccessible Island Flightless Rail, the smallest flightless bird on earth. Its looks like a miniature Kiwi, its feathers resembling hairs rather than real feathers. Ecologically, it lives more like a mouse than a bird, running through narrow tunnels under the grass. It even squeeks like a mouse.

The fourth island is Gough, roughly 400 km southeast of Tristan. At fourty degrees south, it touches the roaring forties. Consequently, it has a much windier and wetter climate than Tristan, with more than twice the amount of rainfall (over 3000 mm). Gough lies near the Subtropical Convergence (STC), a sharp demarcation line between the subtropical and subantarctic ocean, where the water temperature drops abruptly from 18°C to 10°C. The STC is usually at about 42 degrees south, but it moves back and forth a bit, and occasionally runs between Tristan and Gough. Gough is uninhabited, but has a permanently manned weather station, operated by South African meteorologists, who are stationed there for a whole year. Apart from that, Gough is a strict nature reserve, and a UNESCO World Heritage Site, where no landings by visitors are permitted. In the lease contract of the station, between the governments of the UK and South Aftica, it is stipulated that the relief ship, visiting once a year, also has to service Tristan with passengers and cargo. That ship is today the only regular service between Tristan and the mainland, the only other transport being the fishing vessels, or an occasionally visiting cruise ship or private yacht.

Gough has a surface of 65 square km, and, like Tristan and Inaccessible, has sheer cliffs all around, with the exception of the Glenn, a steep valley that runs from the shore inland. The highest point is Edinburgh Peak, at 910 m, almost always in the clouds. Gough has virtually the entire world population of the Tristan Albatross, breeding on the high plains. Like Inaccessible and Nightingale, it has millions of seabirds of many species, and, being larger, it is one of the richest seabird islands in the world.

Gough is the most northerly breeding station in the world for the Giant Petrel. Giant Petrels, formerly believed to be one species, have been split into two, the Southern Giant Petrel with a more southerly distribution in Antarctica and various antarctic and subantarctic islands, and the Northern Giant Petrel with a more northerly distribution, only on subantarctic islands. Their ranges greatly overlap, and in South Georgia for instance, both species breed side by side. In South Georgia it was discovered that they were not one, but two species, which are very hard to ditinguish from each other, but differ in their breeding habits. Strangely enough, the species nesting on Gough is the southern one, not the northern, as one may expect.

Gough has two endemic landbirds, the Gough Island Bunting (not closely related to the buntings of Tristan, Nightingale and Inaccessible), and, of course, the famous flightless Gough Moorhen.

Of the seabirds, the Tristan Albatross, the Atlantic Yellow-nosed Albatross, the Atlantic Petrel, and the Spectacled Petrel are endemic to the Tristan archipelago. The Greater Shearwater is near-endemic, with small numbers also occurring in the Falklands.

Apart from the breeding birds, the islands are frequently visited by vagrants, in a great variety of species, most often blown accross the ocean from the Americas. I have seen a Cattle Egret and a Barn Swallow at Tristan, and Franklin's Gull and Semipalmated Sandpiper at Gough. The most common guest, sometimes a dozen or more at a time, is the South American Purple Gallinule, closely allied to Moorhens. The islanders know it, and call it Guttersnake. The maximum count was 47, in and around the Settlement. They never managed to survive long, and establish themselves as a new inhabitant, but it clearly shows how a large invasion of Moorhens may have occurred in a distant past, colonising Tristan and Gough.

The invertebrate fauna is also interesting, with endemic tiny land snails, and very strange flightless moths. Both of them occur in a variety of species, some confined to one island, others are found on two or more islands, which seems just as unlikely as the flightless Moorhen living on both Gough and Tristan.

4. Early shipping around Tristan

In this chapter I limit myself to shipping up to 1800. After that, there are too many ships. Others keep lists of all the ships up to the present, usually stamp collectors, like Smith (1986, 1991, 1994) and Taylor (2001-2008).

The time up to 1800 can conveniently be divided into three centuries: the 16th, 17th, and 18th century, which can be called the Portuguese, Dutch, and English era, respectively. In the 16th century, there were only Portuguese ships near Tristan, on their way to India and the East Indies. In the 17th century it was almost exclusively the Dutch, when they took over the eastern trade from the Portuguese. In the 18th century the British replaced the Dutch as rulers of the southern oceans.

The English era is well documented. French were also prominent in those days, as were the American whalers and sealers later in the century. Our knowledge of the Dutch era is mainly based on the work of Brander (1940). Being Dutch himself, he had relatively easy access to the Colonial Archives in The Hague. Since Brander's days, access to these archives (no longer colonial) has become much easier, so, being Dutch too, I could find some interesting additions for my earlier book (1997), which I will relate below.

Brander thought he was the first to write about the early Dutch voyages (he says so in his introduction), but it appears that the Italian polar geographer Arnaldo Faustini mentioned them much earlier, around 1925, in his *Annals of Tristan da Cunha*. The *Annals* had not yet been published when he died in 1944. His daughter found the manuscript in the attic after her mother's death, in 1990. She had it transcribed into English, and published it on the internet in the early 2000's, where it can be downloaded as a PDF document (see URL in reference list). Faustini's handwriting must have been pretty awful, and the transcriber probably had a hard time, names of ships and captains often being corrupted beyond recognition.

The Portuguese era really is the great blank in Tristan's shipping history. Most writers only know the voyage of Tristão da Cunha, who discovered the island in 1506. Smith (1991) mentions a Portuguese fleet, visiting Tristan in 1583, and Faustini adds two voyages, in 1520 and 1557. There must have been more. One indication is that in a peat core from Tristan, pollen were found of Ribwort Plantain *Plantago lanceolata*, going back to about 1570 (Ljung & Björk 2011). This is an invasive European species, introduced worldwide by people with dirty feet, or by bringing domestic animals with seeds in their pelts. The early Portuguese had the habit of intruducing goats to islands they found, as a food source for future visitors, and around 1570 only the Portuguese had been near Tristan. None of the visitors (the ones we know of) in the 17th and 18th centuries ever mentions seeing goats, but they probably did not venture far enough inland to encounter them. In my view, they have been there all the time, but did not like the dense, almost impenetrable bush at lower levels and retreated to the higher ground on the mountain, with more open vegetation. The first visitor mentioning goats is Patten, who stayed six months on the island in 1790 (see below). I think they were of Portuguese origin going back to the mid-sixteenth century.

A Portuguese colleague drew my attention to some old history books, and via him I came into contact with naval historian José Manuel Malhão Pereira, who wrote a thesis about early Portuguese voyages (Malhão Pereira 2001). We exchanged dozens of emails, and he overloaded me with digitised old maps, *roteiros* (pilot books), and other publications. His thesis already gave me one or two new voyages. His most spectacular find is a shipwreck as early as 1508, just two years after the island was first seen!

I learned a lot about the early Portuguese navigators. After Bartolomeu Diaz rounded the Cape of Good Hope in 1488, Vasco da Gama reached India in 1498, Pedro Alvares Cabral discovered Brazil in 1500, and Tristão da

Cunha found Tristan in 1506, their scientists and pilots soon started to understand the global circulation patterns of air and sea in the Atlantic Ocean. It took Da Gama nine months to reach India. Cabral did it in six, in spite of his Brazilian detour. Others soon followed his course, using the trade winds, from the Cape Verde Islands to the easternmost point of Brazil. Then they would sail south, and turn southeast towards Martin Vaz and Tristan da Cunha, to reach the Cape, making optimal use of prevailing winds and currents. On the return voyage, they would simply sail in a straight line from the Cape to Saint Helena, using the southeast trade winds. Then they would again cross the Atlantic in a wide arc far to the west, and finally return to Portugal with the westerlies, via the Azores. Thus, their sailing itineraries in the Atlantic would describe a huge figure of eight (just like the Greater Shearwaters of Nightingale - aren't they clever too!). In their *roteiros*, the pilots advised to stay well north of Tristan da Cunha, because of the risk of strong gales further south, so most ships would not go further south than 35 degrees, hundreds of kilometres further north than Tristan. That may be a reason why we see so few Portuguese sightings of the island.

The Portuguese already knew how to use the declination of the compass to estimate their longitude. The declination of the compass is the angle between the true north and the needle pointing at the magnetic north pole (somewhere in northern Canada). Obviously, this angle depends on where you are, relative to these two poles. The magnetic north and south pole are not opposite each other on an axis through the centre of the world. They both lie more or less on the same side of the earth. The lines between them, connecting points with the same declination, called isgones, show all sorts of strange curves, in some cases even forming completely inexplicable (for me...) loops. Also, as the magnetic poles are moving, the pattern changes over time. In the Portuguese era, the line with zero declination (the agonic line) ran north-south through South Africa. At the Cape, the declination was about three degrees east, near Tristan it was almost twenty degrees east. The Portuguese sailors knew this. As the isogones ran parallel to the coast of South Africa, the declination gave them a fair idea of the distance to the Cape. Ships would often note in their journal that they passed Tristan da Cunha *norte-sul* (north-south), based on the observed declination, usually still far to the north of the island.

Today, the isogonic pattern in the South Atlantic is completely different. The agonic line moved westward, passed Tristan around 1730, and now lies much further to the west. The declination at Tristan is now about twenty degrees west, instead of east! Near South Africa, the isogones no longer run parallel to the coast, but hit the shore perpendicular. So today, they would be useless for sailors for establishing their longitude. Fortunately we have other methods now. But I greatly admire the old Portuguese sailors for already knowing all this!

Portuguese captains were also asked to note all the animals and plants they would see, which they called *sinais* (signs or signals), such as birds, floating seaweed, driftwood, smell and colour of the sea, and so on. In European waters they knew birds were a sure sign of nearby land, and in the south they thought so too. Gannets and gull species certainly indicate a nearby coast, but petrels and albatrosses do not. They roam widely, all over he ocean, New Zealand species frequenting the South Atlantic, and vice versa. When approaching Tristan waters, they often noted *corvas pretas, bicos brancas*, black ravens with white beaks. These must have been White-chinned Petrels (*Procellaria aequinoctialis*), which are very common in the Southern Ocean, and often follow ships in their wake. Near Tristan they could even have been the closely allied Spectacled Petrels (*P. conspicillata*), endemic to Inaccessible Island. They also saw black-and-white checkered birds, which they called *feijãos*, speckled beans, a very strange name for a bird.

They must have been Cape Petrels (*Daption capense*), which also like to follow ships. Seeing those was no indication of nearby islands at all, they just appeared because the ships approached the nutrient rich subantarcic seas. Feeding frenzies of these birds can be seen anywhere between Cape Horn and the Cape of Good Hope, wherever they find a local patch of krill. The nearest breeding stations of the *feijãos* are South Georgia, the South Sandwich Islands, and Bouvetøya, more than thousand kilometres further south. Floating seaweed is also an unreliable sign. On one of my South Atlantic voyages we saw floating seaweed all the way between Tristan and Saint Helena.

Let's turn our attention to Tristan visitors.

In 1505 Gonçalo Alvares, sailing as a pilot with Pero d'Anhaya, was blown too far to the south. On his way to the Cape he saw a godforsaken island which was named after him, Ilha de Gonçalo Alvares, which was later rediscovered as Gough Island. Gonçalo Alvares was erroneously changed into Diego Alvarez (Spanish spelling). According to Uhden (1939) this happened as late as the early eighteenth century, as he found the name Conçalo Alvares or Alvarez on the maps of Diego Ribeiro (1529), Desceliers (1546), Gastaldi (1562), Jansonius (1560), and d'Isle (1700). I found some more in maps dated 1519, 1539, 1544, and 1570, but in a map of 1560 I already saw the name Diogo Alvares (Portuguese spelling). The other maps often showed the name abbreviated to Ilha de go Alvares. De go could easily be misread as Diogo. People have suggested it would be nice to change the name of Gough back into the old name Diego Alvarez. But that is not the old name. It should be Gonçalo Alvares. Or, to avoid the confusion about Diogo, Diego, de go, or Gonçalo, let's just call it Alvares Island, instead of the ugly-sounding Gough.

On the early maps, Alvares Island is drawn a bit too far to the north, at 39 degrees instead of 40, which later confused Dutch sailors. Because of the name change, Gonçalo Alvares' discovery is often given as a 'maybe', but I think we are now confident enough to remove the question mark.

The story of Tristão da Cunha finding the islands in 1506, which were named after him is well known, but needs some clarifications. According to MacKay (1963) Tristão's ship was called *Capitão Mor*. This is an error, which has been repeated by other authors, including myself, I am afraid (Beintema 1997). Capitão-mor was not his ship, but his title. His ship was the *Espirito Santo*. In a fleet, every ship had her own captain, but only one of them would be the chief captain, or captain-major: capitão-mor. This was not a permanent title, like Admiral, but just an assignment for the voyage. The size of the fleet varies between sources: 14, 15, 16 or just 6. I think they sailed from Lisbon to Brazil with 16 ships. One was sent back to Portugal with sick men, one ship was lost, and 14 reached the waters around Tristan da Cunha. Da Cunha sailed with Afonso de Alboquerque (or Dalboquerque), future viceroy of India. D'Alboquerque commanded a sub-fleet of six ships, with fourhundred men, with a special assignment to control the Moors on the African eastcoast, so he was a capitão-mor too, and captain of the *Flor de la Mar*. Having two chief captains on one fleet is asking for trouble, and indeed, there were a lot of disputes between the two. D'Alboquerque was especially irritated because da Cunha's ship was slow, and they often had to wait for each other. It was already late in the season and it was questionable whether they would still be able to safely round the Cape. After discovering Tristan, the weather took the matter in hand. In a heavy gale the fleet was dispersed, and the ships arrived on the east coast of Africa at different times. Since 1509, the charts show seven little dots, with the remark *'Ilhas que achou Tristão da Cunha'* - islands found by Tristão da Cunha. Malhão Pereira found a very nice drawing, in colour, of the entire fleet of fourteen near Tristan, with small boats filled with rowing men, exploring the islands.

We sometimes see Tristão spelled as Tristam. Note that ão and am are pronounced the same in Portuguese (aaung, with a nasal ending, like in the French bon). Similarly, we often see Sam instead of São (Saint).

We all agree that Tristão is the one who discovered Tristan. But was he really the first to see the islands? In June 1503 the French adventurer Binot de Paulmier de Gonneville left the Harbour of Honfleur with his ship *L'Espoir*. He heard of the riches brought back by Vasco da Gama and Cabral and wanted his share, ingoring the fact that the route to India was forbidden for non-Portuguese ships. Columbus, Vasco da Gama, Cabral and Tristão da Cunha all travelled with large fleets. De Gonneville sailed solo. On November 9th, they encountered large fields of floating seaweed, with roots. They first thought that it meant they were close to the Cape, but because they never saw Gannets, which are always seen there, they figured it must have been somewhere else. They also found it was suddenly much colder. Then *L'Espoir* had strong contrary winds for several weeks, and they had to reef to prevent being blown back to the north. They drifted in all directions and were utterly lost. The crew was suffering from scurvy, and several men had already died, including the all important Portuguese pilot. What made them decide to travel west in the end, instead of east, is totally unclear. Eventually, they ended up on the coast of southern Brazil, where they met the same kind of friendly natives Cabral had seen. They had found paradise, and stayed there for six months, having a wonderful time, then gradually moved north along the coast, to return to France in 1505. De Gonneville had no idea he had been in South America, and claimed to have found the fabled Great South Land that everybody was hoping to find. He called it *Les Indes Méridionales*, the South Indies. His stories became a myth, especially in the 17th and 18th centuries, and were the basis for several French expeditions to find the Great South Land, leading to the discoveries of Bouvet in the

South Atlantic, and Kerguelen in the southern Indian Ocean. De Gonnevilles log was lost. It surfaced as late as 1869, and was made available for scientists. The log was thoroughly analysed and historians could even identify the various places where De Gonneville had been ashore (Perrone-Moisés 1992, Bueno 1998). They concluded that the seaweed must have been near Tristan da Cunha, and one of them even says that the island was seen. In the version of the log I have read, there is no mention of seeing an island at all, only the seaweed, which could have been anywhere. So in my judgment, de Gonneville never saw Tristan.

In April 1508, a fleet of seventeen sails departed for India, divided into two captaincies. Jorge da Guiar was capitão-mor of thirteen ships, eight of which would go for spices, and five would stay on the Arabian coast. The remaining four vessels were to explore Malacca, under command of Diogo Lopes de Segueira. Jorge da Guiar lost his ship *Sam João* during the night on the shores of Tristan da Cunha. There are beautiful drawings of the accident, in colour, in two different Portuguese pilot books, clearly made by the same artist who produced the drawing of Tristão's explorations (Soeiro de Brito et al. 1992).

In his thesis, Malhao Pereira (2001) plotted the itineraries of various voyages. Most of them stayed well north of Tristan, but in 1535 Fernão Peres de Andrade, with the *Espera*, came very close, and indeed the text says he saw the largest of the islands, on June 17th. He found the island 'well shadowed' and supposed it was not very high, which probably means the mountain was hidden in clouds. Two years later, in 1537, André Vaz (no ship's name mentioned) passed the island even closer, but did not mention seeing it. Perhaps clouds or fog blocked his view, or perhaps he saw it without saying so. But as he was so close, I add him to my list with a question mark.

For the 16th century, Faustini gives us two more Portuguese visitors, both questionable: in 1520 the navigator Ruy Vaz Pereira, captain of the ship *Las Rafael*, called at Tristan for water, on his way to Muscat. And in 1557 Luis Fernandez de Vasemcellos, captain of the ship *Santa Maria de Carca*, sailed with the new Portuguese governor from the Canary Islands to Brazil, and then on to Tristan da Cunha, which he sighted at the beginning of July.

Rui Vaz Pereira's ship's name was *Sam Rafael* (misspelled by Faustini's transcriber). We could not find any mention of Tristan. All stories about him are about catching a giant fish, almost the size of his ship, near the Cape. It is a complete mystery how Faustini got the information about fetching water and sailing to Muscat. Unfortunately, the list of references has not been included in the transcribed *Annals*, so we have no idea where Faustini found this information. We keep this voyage on the list, with a huge question mark.

Regarding the 1557 voyage: the captain's name was Vasconcelos, and his ship the *Santa Maria da Craça*. Malhão Pereira found the voyage, but there is no indication that Vasconcelos saw the island. So here is the next question mark. Finally, there is the 1583 fleet, mentioned by Smith (1991). He refers to an earlier listing made by Butler in 1952. I have not been able to find this document, and even my English, stamp-collecting Tristan friends had never heard of it. They knew Butler was a philatelist too, who even designed new stamps for Tristan which were never produced. We could not find a 1583 voyage. This leaves us with another big question mark. That concludes the Portuguese era. Lots of uncertainties, and no landing with goats could be identified.

A hundred years after the Spanish and the Portuguese conquered the world's oceans, the Dutch entered the naval scene and sailed to the East Indies, where they ultimately replaced the Portuguese. The East Indian Spice Islands would become a Dutch colony. Dutch pilots did their homework. They were a bit bolder, or perhaps more reckless that the Portuguese, and ventured further south to make better use of the strong westerlies. Following that course, Tristan da Cunha appears more regularly in shipping reports.

According to Faustini, the second Dutch voyage to the East Indies, under Jacob van Neck (spelled Heck by his transcriber) sailed close to the northern shores of Tristan. It was on this voyage, that the uninhabited island of Mauritius in the Indian Ocean was discovered and named after Prince Maurits. The crew found prolific wildlife. They rode on the backs of giant tortoises and killed the first Dodos, the most famous of all extinct animals. The journals of this voyage have been republished by the Van Linschoten Society in five volumes, from 1938 to 1949. I read all the reports of the ships involved, and none of them mentions seeing Tristan da Cunha. They all mention large numbers of birds at a latitude of 34 degrees south, which made them think they might be nearby, but they were still a few hundred kilometres north of the island. They saw albatrosses, Cape Pigeons, and probably the endemic Spectacled Petrels from Inaccessible. And large numbers of birds like Turtle Doves, which most likely were Broad-billed Prions. They just ran into a rich food patch which had attracted the birds, but did not see Tristan. They stayed at 34 degrees until they reached the Cape.

The first non-Portuguese ship, the Dutch *Bruinvis*, skipper Willem van Westzanen, in a fleet commanded by Jacob van Heemskerk, arrived at Tristan in 1601. Van Heemskerk was famous for sailing with Willem Barentsz in 1596, discovering Spitsbergen, and wintering on Nova Zembla in a failed attempt to find the NE-Passage to China and the Spice Islands. Willem van Westzanen saw land during the night, and sailed at a safe distance until morning. Then they returned to the island, saw the impressive cliffs and snow on the peak, found no safe anchorage, and when sudden wind gusts came rolling down from the mountain, they fled for safer water.

In 1610, allegedly, the first sighting of Tristan by a British ship, the *Globe*, took place.

I think this is a wrong interpretation of the journal. The ship's log mentions passing the Abrolhos near the Brazilian coast on April 14th. The next entry is May 14th, when they had reached 34 degrees south. On May 18th they saw the Cape of Good Hope. That means that on the 14th, they must have been closer to the Cape than to Tristan, as it usually took about two weeks to sail the distance from Tristan to the Cape. From the 14th till the 17th they had a severe storm from the west. They reefed, fearing that they would be blown onto the coast, which must have been Africa, not Tristan. They found the compass declination to be only three degrees, indicating they were near the Cape, but figured they were a bit farther away because of the birds they saw: *for wee saw dyvers foules that keepe aboute the Cape, etc. These followed us from the island of Tristan d'a Chuna to the Cape.* The log does not mention an actual sighting of the island, which they must have passed north of 34 degrees, more than 300 km away. It is quite normal to see large numbers of petrels that breed in the Tristan archipelago in the waters between Tristan and the Cape, but that does not mean you are close to land at all (see also van Neck). So I don't believe they saw the island. The name was only used, together with the Cape, to indicate the huge ocean area between the two places. Minor detail: the voyage took place in 1611, not 1610.

According to Brander, the Dutchman Willem IJsbrantsz. Bontekoe saw Tristan in 1618, but Bontekoe's report explicitly states that they must have passed close, but did not see the island. Faustini also mentions Bontekoe (spelled Brutekoé), but for the year 1619.

Faustini names three French ships in 1620, *Montmorancy*, *Esperance*, and *Hermitage*, skipper Beaulieu. They tried to approach the island but found their way barred by large fields of floating seaweed.

Smith (1991) mentions a Dutch fleet and an attempted landing in 1626. I found no trace of this visit in any other source, and since Smith does not give a source for this one (which he otherwise does), and the failed landing went exactly like Speckx's attempt that according to Brander took place in 1628, I conclude that this is a duplication, 1628 being misread or misprinted as 1626. I checked the visit of Jacob Speckx, where Brander does not name a ship. In fact, Speckx visited Tristan in 1629, not 1628. He left Texel in December 1628, and arrived at Tristan on 7 June, 1629, with the ships *Hollandia, Der Goes, Oostzaenen*, and *Westzaenen*. An attempted landing failed. Admiral Speckx was appointed Governor General of the Dutch East Indies, after Jan Pietersz. Coen, the first Governor General, died. When digging into Speckx, I found a 'true new' visitor: Artus Gijsels, who was sent as governor to Ambon, with the *Deventer, Middelburg*, and *Hof van Holland*. On 3 August 1630 he saw the islands of Tristan. Like Speckx the year before, he was impressed by their beauty and ruggedness, but he did not attempt a landing.

The first documented landing on Tristan took place in February 1643, with the Dutch ship *Heemstede*, skipper Claes Gerritszoon Bierenbroodspot, from the city of Hoorn. Extensive reports on this visit, and the subsequent Dutch expeditions later in the 17th century, are found in several books and publications on Tristan, so here I keep it brief.

The *Heemstede* found the anchorage opposite the waterfall, and stayed there for eight days in fine weather. People went ashore several times, found the drinking water of excellent quality, and praised the taste of the wild celery they found. The beach was loaded with Elephant Seals and Fur Seals which they killed as often as they could, bashing in their heads. Their report recommends that it would be advisable for the East India Company to claim the island as a refreshment post as it lay so perfectly halfway between Holland and the East Indies.

In 1652 the Cape Province became a Dutch colony, and the governor, Jan van Riebeeck, organised an exploratory mission to Tristan da Cunha. So, in November 1655 the galiot *'t Nachtglas*, skipper Jan Jacobz, left the Cape for

Tristan, the first true Tristan expedition. *'t Nachtglas* stayed for eight days at Tristan, in January 1656. They first landed on Inaccessible, which was unnamed yet, and named it Nachtglas Eylandt. They also visited Nightingale, which they called Gebroocken Eylandt (Broken Island). They tried to land at several places around Tristan, but found it was only safe to land near the same waterfall, where they found a plaque, nailed to the rocks by Bierenbroodspot in 1643. Wherever they landed, they had to kill seals left and right, to create passage. The report from *'t Nachtglas* states that generally speaking, the waters around the island are too dangerous. The usefulness of Tristan as a station for the company would be doubtful. No further steps were undertaken.

In 1659, the Dutch ship *Graveland* visited Tristan. They made an easy landing near the waterfall, and reported favourably about the island. Thus, the interest in the island was renewed. In 1665 the *Pimpel* was sent to explore and compare the usefulness of Martin Vaz, Tristan da Cunha, and Diego Alvarez. The *Pimpel* did not get any further than Martin Vaz, and never reached Tristan. In 1669, The *Grundel*, captain Gerritsz Riddermuis, visited Martin Vaz and Tristan with the same purpose. Their conclusion was that, with favourable winds, small ships could get close enough to go ashore for refreshments, but that this would be impossible for larger ships because there was no safe anchorage.

A 'new' visitor to Gough comes from the report of the *Grundel*. After visiting Tristan they searched for Diego Alvarez in vain, at the 38th parallel. Therefore, the captain concluded that twelve years earlier Rijklof van Goens, who saw an island at 40 degrees, which nobody believed could have been Diego Alvarez, must have seen that island after all! So I dug out Van Goens' report. With his ship *Orangie* he circled around the island in February 1657. Diego Alvarez was at that time believed to lie further north, so Van Goens did not know what he was seeing. Meanwhile, Gough has been moved to the right place on the map, and, judging from his descriptions, Van Goens cannot have seen anything else.

Faustini mentions a French fleet sailing along the southern coast of Tristan in November 1666, under the command of Marquis de Mondevergne. They observed: 'The main island has a peak higher as that of Teneriffe', which of course is not true as we now know. The Pico the Teide is almost twice as high as Tristan, and has long been believed to be the highest mountain on earth. In 1676 there was another French ship, the *Vautour* whose crew observed: 'We found three islands, one large and two small', and 'The peak is covered with snow. The island is uninhabited'.

Regarding the early Dutch visits, most Tristan books or ships' listings rely heavily on Jan Brander (1940). Brander, a Dutch teacher in geography and history, did a lot of digging in the archives of the Dutch East India Company, and since most Tristan authors cannot read Dutch, nobody followed him there. I did. This was a rewarding exercise, because today these archives are much easier to access than in Brander's days. An important source for instance is *Dutch Asiatic Shipping*, by Bruijn et al. (1979), which sums op over 4700 outgoing voyages of the Dutch East India Company. It mentions the following hitherto 'unknown' visits to Tristan:

1646: *Witte Olifant*, skipper Klaas Bot,
1646: *Koning David*, skipper Reinier Egbertz,
1646: *Witte Paard*,
1658: *Elburg*, skipper Pereboom,
1681: *Ternate*, skipper Jan Gerritz.

These visits are also mentioned by Headland (1992), with reference to Bruijn et al. However, I discovered that the first four never visited Tristan, so Brander overlooked less than I at first thought.

On April 6th, 1646, the *Witte Olifant, Koning David* and

Witte Paard left Texel. On September 5th they are said to have arrived at Tristan. From archives kept in Batavia, it appears that the story is different. The three ships sailed in the company of two others, the *Zeelandia* and the *Patria*. On the 5th of September, 'about Tristan da Cunha', they had to split up, because the first three ships ran out of drinking water and had to divert to the Cape of Good Hope. *Zeelandia* and *Patria* arrived in Batavia on November 12th, the three others in December. The captains were reprimanded, because they would have been on schedule, had they taken more water at the Cape Verde Islands. From this story I must conclude that none of these ships reached Tristan. They did not mention actually seeing the island, and the name Tristan only appears in that one sentence about splitting up.

The *Elburg* was said to have been forced to spend four weeks at Tristan, because of ... calms. Who can believe that? The diary of Jan van Riebeeck, Governor of the Cape, says that the ship arrived at the Cape on 13 April 1658, that no place was visited on the way, and that the ship had been troubled by calms for four weeks, between Tristan da Cunha and the Cape. So this is another non-visitor.

Remains the *Ternate*, which touched Tristan indeed: it nearly wrecked. The *Ternate* left Texel on 31 May 1681. In thick fog it scraped the rocks of Tristan and sprang a small leak. Fortunately it was able to get away and proceed to the Cape, where it arrived on 27 September. The *Ternate* continued its voyage to Batavia, but never sailed again.

There is a wonderful story about the expedition of the *Tonquin Merchant* in 1684. British colonists in Saint Helena were looking to expand, and wanted to colonise Tristan da Cunha. People interested were promised free passage, a salary, and free supplies. According to Faustini, the ship sailed to Tristan. Captain Knox was ashore, but when he wanted to go back on board, the crew cut the cables and sailed away, leaving him behind. However, the real story is different. The incident of the crew leaving the captain behind did take place but not at Tristan. It happened in Saint Helena when the expedition was about to leave. So the expedition never took place, and the disappointed Captain Knox had to return empty-handed to England from Saint Helena.

There are two more British voyages in the 17th century. There was a 'maybe' landing on Gough by Antoine de la Roche in 1675, and the visit of the *Welfare*, *Kent* and *Rainbow* in 1685.

There was another Dutch visit which we do not often see in Tristan literature. In 1690 the French Huguenot François Leguat (after fleeing from France) sailed from Texel with the *Hirondelle* (in English sources named *Swallow*) to Rodriguez in the Indian Ocean. He saw Tristan, but a landing could not be made. Leguat became famous for his reports on now extinct birds on Rodriguez (e.g. the Dodo-like Solitaire).

The last Dutch expedition took place in 1696. Three ships left Texel, the frigate *Geelvinck*, the hooker *Nijptang*, and the galiot *'t Weseltje*, under the command of Willem de Vlamingh. They were asked to look for the ship *Ridderschap* which had vanished after leaving the Cape, in early 1694, and to investigate Tristan da Cunha once more. The ships lost each other in bad weather, and did not arrive at Tristan together. Only the Geelvinck managed to make a landing on Nightingale, and their judgment was that, especially in winter, (they were there in August) the islands were pretty useless, because of strong winds, mist, and generally bad weather. After Tristan, they visited Amsterdam Island and St.Paul in the Indian Ocean, and eventually they reached the west coast of Australia, thinking it was the Great South Land at last!

The Tasmanian historian Irene Schaffer wrote *The Sea shall not have them* (2010), about the Tasmanian descendants of Stephen and Peggy White, survivors of the famous *Blenden Hall* shipwreck at Inaccessible Island (see chapter 8). In an appendix about Tristan da Cunha she mentions a ship I had never heard of, the *Vlaming*, with

Captain Francis Cheyne, visiting Tristan in 1697. This is an error. When England wanted to establish a convict colony in Australia in the second half of the 18th century, Alexander Dalrymple, hydrographer for the East India Company, suggested using Tristan da Cunha instead, as it was closer and guards would not be needed. In 1786 he wrote the pamphlet *A serious Admonition to the Publick on the intended Thief Colony at Botany Bay*, later re-published by George Mackeness in his series of Australian Historical Monographs (1943). Dalrymple added a description of Tristan da Cunha, which Schaffer quoted. I found Dalrymple's text on the internet (see URL in reference list). He took his description from the *English Pilot for Oriental Navigation* which has a footnote at the end, saying that the 'Burgomaster' (whoever that may have been) told him of Vlaming visiting the island in 1697. On the next page there is another description of Tristan from the journal of Captain Francis Cheyne. The footnote about Vlaming continues at the bottom of this page, below Cheyne's text. Schaffer mistook this as Vlaming being Cheyne's ship. The footnote actually refers to the visit of Willem de Vlamingh. In Cheyne's report, there is no name of a ship mentioned, or a year of his visit.

The text Dalrymple quoted from the *English Pilot* mentions 'a strange bird that goes upright'. Perhaps this is the first description of an Island Cock?

The 18th century begins with Edmund Halley (yes, the famous one who had a comet named after him), who sailed close by with the *Paramore* in 1700. He attempted no landing. He saw Nightingale and Inaccessible at close range, but Tristan was hidden in clouds. Only when he was 26 miles east of Tristan did the peak suddenly appear above the clouds.

Headland (1992) mentions the French ship *St. Louis*, passing Tristan in 1708, on the way from Cape Horn to the Cape of Good Hope. I am not sure whether they really saw the island, so I leave this one with a question mark.

In 1732 the *Richmond* visited Diego Alvarez. Captain Gough did not know the island already had a name, so he named it after himself. Unfortunately the name stuck. The same happened with Gebroocken Eylant (Broken Island), which was visited by ignorant Captain Gamiel Nightingale, who named it after himself in 1760. And in 1767 the French Captain d'Etchevery renamed Nachtglas Island and called it Inaccessible, which is pronounced differently in French but spelled the same as in English. Not many people know that Inaccessible today actually still has a French name!

Margareth McKay, in her book *The Angry Island* (1963), mentions a visit around 1775 of the French Captain D'Après de Mannevillette, who was the only one ever reporting large numbers of sea turtles on Tristan beaches. Being a biologist, I know that this is impossible. D'Après de Mannevillette wrote a description of the sea route to the East (*Le Neptune Oriental*, 1745, second imprint 1775), based on various ships' journals, and never visited Tristan himself. Initially I thought he had just mixed up reports from other islands, but later I found another explanation for the turtles. D'Après de Mannevillette and Alexander Dalrymple were friends and exchanged sources. When I saw Dalrymple's description of Tristan with his strange upright bird, it reminded me of a similar bird in D'Après de Mannevillette's text. Comparing the two texts, they appeared to be virtually identical, word by word, except that in Dalrymples version the beaches were crowded with seals instead of turtles. So D'Après de Mannevillette did not mix up reports. The turtles were just a stupid translation error. It is a bit mysterious, though, that while some of Dalrymple's seals were as big as elephants, the French turtles were as big as 'veaux marins', sea cows (Lamantins), while the French word for elephant is just éléphant, as in English.

In the French translation the *English Pilot* is quoted as *le Pilote Anglois*, which in the Dutch translation of the French became 'Engelsche Stuurman', English Mate, changing the book into a person. The upright bird became 'oiseau qui marche perpendiculairement', bird that walks perpendicular. In the Dutch translation, it had evolved into a strange bird going straight up and down. The Dutch version was the first I saw, so that bird puzzled me, but now I understand where it came from. I really think it was an Island Cock.

D'Après de Mannevillette also quotes from the journals of *l'Adelaide*, *l'Eclatant*, and *le Fendant*, which visited Tristan in 1712, and escaped the notice of previous Tristan historians. I thought I was the first to discover them, but it appears that Faustini mentions them too (when I wrote my book in 1997, Faustini had not yet surfaced). D'Après de Mannevillette adds a final report from a fourth ship, *Le Rouillé*, which sailed around Tristan in 1755.

Austria has never been known as a seagoing nation, establishing overseas colonies. Yet, Tristan da Cunha has been Austrian territory. In 1775 the *Société Impériale Asiatique de Trieste* was founded in Antwerp. In those days, Belgium was Austrian territory, ruled by Empress Maria Theresa. The firm was financed by Count Proli from Antwerp, and Willem Bolts from Amsterdam was appointed director. In 1775 Bolts sailed with the *Joseph et Thérèse* from London to Asia. They saw Tristan, and decided to appropriate it for the Austrian Empress. Business did not go well, so in 1781 the partners in the firm had to avoid bankruptcy by selling all their personal assets, which included non-existing real estate on Tristan. In 1785 the firm went down, but the island remained Austrian property. In all likelihood, Emperor Joseph II, who succeeded Maria Theresa in 1780, never knew this.

Towards the end of the 18th century shipping around Tristan intensified, everybody wanting his share in oil and fur from whales and seals. Tristan got its first inhabitant, the American John Patten, captain of the *Industry*. Between August 1790 and April 1791 he killed 5600 Fur Seals. At the same time, Captain Colquhoun of the *Betsy* planted potatoes. One of the many American whalers around Tristan was the *Grand Turk*, which visited Tristan in 1792. On board was the eccentric Jonathan Lambert, who would return almost twenty years later to establish his private kingdom.

1792 is also the year the first British landing took place. The *Lion*, *Jackal* and *Hindostan* were on their way to China, to deliver the first British Ambassador. They landed near the waterfall to get fresh drinking water. This is also the time the first serious biological research on the island took place. The French botanist Aubert du Petit Thouars, with the ship *Le Courier*, captain Gars, spent a couple of days ashore in 1793 collecting plants. He found many new species, several of which were named after him. He tried to climb the mountain, but halfway he found his way blocked by steep cliffs, and had to return. He did not make it in daylight and had to spend the night in the wild, soaked by rain.

Finally I managed to find a ship that had not been listed before by Tristan writers, not even by Faustini, in *The Oriental Navigator* (Purdy 1816), which is largely based on the *Neptune Oriental* by D'Après de Mannevillette. In 1795 the *Providence*, Captain Broughton visited Tristan.

The French naturalist du Petit Thouars saw albatrosses, but he does not mention Moorhens. Gisela Eber used this as an argument for her conclusion that there never was a Moorhen on Tristan. John Patten (1790) on the other hand, knew them very well. He wrote (quoted by Purdy 1816 and Morrell 1832):

"Of birds, the principal were a kind of gannets, like wild geese, which the sailors considered excellent food; penguins, albatrosses, Cape cocks and hens, and a bird like a partridge, but of a black colour, which cannot fly, is easily run down, and is very well flavoured; and a variety of small birds that frequent the bushes and underwood. Abundance of birds' eggs are to be obtained in the proper season."

There is no doubt that Patten consumed the extinct flightless Moorhen of Tristan da Cunha.

5. King Jonathan

The year 1810 is a turning point in Tristan history. On September 27th, three men came to Tristan, and although none of them produced offspring, the island has been permanently inhabited ever since. The modern history of Tristan began at that time.

The three were an Italian, Tomaso Corri (later Anglicised to Thomas Currie, also spelled Curry), and two Americans. One was probably a criminal named Williams, who was calling himself Millet. The other was Jonathan Lambert of Massachusetts, who had visited the island earlier on board the *Grand Turk*. On February 4th 1811 Lambert crowned himself King of the Kingdom of the Refreshment Islands. Tristan was re-named Refreshment Island, Inaccessible became Printard Island, and Nightingale Lovel Island. The King ruled a nation consisting of two people. His proclamation was published in the Boston Gazette of July 18th, 1811. The full text has been quoted in many sources so here I only give the first paragraph:

"know all men by these presents, that I, Jonathan Lambert, late of Salem, in the State of Massachusetts, United States of America, and citizen thereof, have this 4th day of February in the year of Our Lord Eighteen hundred and eleven, taken absolute possession of the island of Tristan da Cunha, so called, viz. the great island and the other two known by the names of Inaccessible and Nightingale Islands, solely for myself and my heirs, for ever, with the right of conveying the whole, or any part of thereof, to one or more persons, by deed of sale, free gift or otherwise, as I, or they (my Heirs), may hereafter think fitting or proper."

He also designed a flag, and promised to treat visitors with hospitality and good fellowship, offering trade in refreshments, and so on, and so on. The document was signed by himself and, as a witness, Andrew Millet. Lambert had great plans, writing to his friend Captain Briggs of all his business proposals.

The Kingdom did not last long. On May 17th, 1812, The King, Williams (AKA Millet), and another American who had just joined them, perished at sea when out fishing and collecting wreckage. Thomas Currie was left as sole inhabitant. A year later, he was joined by John Tankard and John Talsen, so there were once more three inhabitants.

During the Anglo-American war of 1812-1814, the Americans used Tristan as a base to capture British vessels, and several skirmishes at sea took place. The Americans often took vegetables or animals without asking the three men, much to their displeasure. Eventually, Tankard and Talsen left Currie, who was then alone again. Later, he got a new companion, Bastiano Poncho, from the Balearic Islands. Poncho agreed to serve Currie for two years, but he departed before this term ended. Maybe, Currie was not such pleasant company.

In August 1816, a British garrison from the Cape occupied Tristan. The official reason was that Napoleon had been exiled to Saint Helena, and the British were afraid the French would use Tristan as a base to free him. That sounds like a ridiculous plan. In terms of distance, they might as well have occupied the entire west coast of Africa and the east coast of Brazil. A second motive could be that, in those days, although the Anglo-American war was over, there was still a great deal of American piracy around Tristan waters. Having a garrison on Tristan, meant that the sailing route to India could be better guarded.

The soldiers interrogated Currie about the disappearance of Lambert, and his stories were sometimes contradictory. As a result, the soldiers formed the impression that Currie might have murdered them. Currie also had stories about a great treasure he had hidden, and substantiated it by showing golden coins every now and then. His money also gave him access to the canteen of the garrison, where he spent most of his time intoxicated. He promised to tell his best friend where his treasure were hidden, so many soldiers treated him to free drinks. In 1817, when all his best friends were still there, he drank himself into oblivion and died. Currie's treasure gave inspiration to Primo Levi for the chapter 'Mercury' in his book *Il sistema periodico* (1975).

In the 1950s, a South African newspaper tycoon found out that Thomas Currie was one of his ancestors, and that this entitled him to lay claim to the treasure. He announced an expedition to Tristan to find it, using the latest techniques. The expedition never happened, and the eruption in 1961 covered all the places where it was likely to be found with several metres of thick lava.

King Jonathan was the second person I know of to see Island Cocks. He wrote to his friend Captain Briggs:

"We have the little black Cock in great numbers, and in the fall (they) are very fat and delicate. We caught some hundreds last year with a dog..."

Again there is no doubt: the King consumed the extinct flightless Moorhen of Tristan da Cunha.

6. Dugald Carmichael

When Emperor Napoleon went out campaigning, he travelled in his personal chariot, with all sorts of fancy things on board. He had a large collection of beautiful uniforms and a fine collection of weapons, including two pistols with mother-of-pearl grips. He had a chest, ornamented with gems and filled with money, crystal caraffes with booze, and, for his personal hygiene, two silver pos, a box of black ivory containing sandalwood tooth picks, and a golden tongue scraper. Napoleon lost his chariot, with its valuable contents, in Waterloo. He was exiled to Saint Helena, and the British garrison arrived at Tristan.

On November 3rd 1816, HMS *Falmouth* left Cape Town, with 50 soldiers, mostly enlisted Hottentots, under the command of Captain Cloete. They also carried a huge quantity of varied equipment and cattle. On board was Captain Dugald Carmichael, who joined the garrison as a naturalist. The voyage was terrible. It took them 25 days to reach Tristan, 13 more than planned (we will see later that this is not yet a record...). They were tossed around so badly in heavy gales that all the cattle, horses and sheep broke their legs and died or had to be euthanised. After the encampment on Tristan was established, the *Falmouth* returned to the Cape, taking Comilla, who had seen enough of Currie.

In May 1817, the British decided that the occupation of Tristan made no sense, and the garrison was withdrawn. They sent the *Conqueror* to pick up the soldiers and their effects, but it was not possible to take everything on board. A small group of men stayed behind, to keep an eye on the remaining material. The commander of Saint Helena, who also supervised things on Ascension, saw an opportunity. There was a terrible shortage of drinking water and materials on Ascension, so he sent the *Julia* to Tristan, to take fresh water and as many of the goods as possible. The

Julia anchored off Tristan on October 1st, and was loaded, but during the night the ship was lifted off the anchor by sudden swells, and thrown on the rocks. Almost thirty men drowned.

On November 17th, the *Eurydice* came to take the last people and materials. Most men were happy to return to sunny Cape Town. The exception was Corporal William Glass from Kelso, Scotland. He loved the climate of Tristan, which reminded him of home. He came to Tristan with his wife and a little child, and they enjoyed their time on the island. His wife was a Cape Coloured woman, Maria Magdalena Leenders, whose father was Dutch. Glass asked permission to remain and occupy the island for the British Crown. Two companions stayed with him, Samuel Burnell, and John Nankivel. The three of them signed a co-partnership agreement, containing the following sentences:

"1st. That the stock and stores of every description in possession of the Firm shall be considered as belonging equally to each. 2nd. That whatever profit may arise from the concern shall be equally divided. 3rd. All purchases to be paid for equally by each. 4th. That in order to ensure the harmony of the Firm, no member shall assume any superiority whatever, but all to be considered as equal in every respect, each performing his proportion of labour, if not prevented by sickness."

Therefore, after having been a Kingdom, Tristan had now become the first communist country in the world. Later this agreement became a sort of constitution for Tristan, the essence of which is still valid today.

Dugald Carmichael had left Tristan in March, so he only stayed during the summer months, with relatively nice weather. Yet he recorded rain every other day, as well as frequent storms, which usually lasted less than a day. He found the rain frustrating. Under the cloud cover he could always see the sun shining on the sea in the distance. That was the effect of the mountain, where the rising air would cause cloud formation and rain.

Carmichael described the vegetation and the rock formations in great detail. He noted that the plateau on the NW side of the island, where the garrison camped, and where earlier adventurers had lived, was covered by a dense, almost impenetrable vegetation of *Phylica* trees with dense undergrowth. Only a small part, near the landing place, had been cleared. He found the black soil of good quality, and saw a great potential for further clearing and agriculture.

On January 4th 1817, Carmichael, together with some friends, was the first to climb the peak. This was quite an achievement, as there were no trails, and fighting your way through the dense vegetation, climbing more than 2000 m from sea level and back is no easy feat. First they had to struggle to get to the foot of the cliffs, and then find a way up through one of the steep ravines. Today there are only three places from where you can climb up that are within walking distance of the Settlement: Pigbite, east of the Settlement, Burntwood at the southern end of the Settlement Plateau, and Hottentot Gulch (named after the Hottentot garrison) just above the Settlement. For the rest, the cliffs are virtually unsurmountable. In all likelihood Carmichael went up through Hottentot Gulch.

With great difficulty they reached the top of the cliff from where they admired the views of distant Nightingale and Inaccessible. From there, it sloped gently upwards, first through wet, soggy swamps, later, where it got steeper, through loose gravel. At the top they found a clear crater lake, with not a single sign of life. Wild, and desolate. On the steep sides of the mountain, they had to balance on narrow ridges, with deep ravines on either side. They had to take great care not to fall down, as they were sometimes almost blown off their feet by sudden wind gusts coming down from the mountain. On those ridges they found many albatrosses nesting or resting.

Albatrosses are great flyers, and can cover thousands of miles without any effort, but just like like real aeroplanes, they need a runway to get airborne. Carmichael saw how they would run downhill, faster and faster, with outstretched wings, until the wind lifted them off the ground. As a scientific experiment he took an albatross by the wingtip and threw it over the edge of the ravine. He watched the bird spiral down, hopelessly flapping its wings, unable to recover flight, eventually falling down like a brick.

During his stay on Tristan, Carmichael collected many plants, fish, and birds. He made a fine collection of seabirds. He also captured thrushes, and the now extinct bunting. About the land birds he writes:

"The only land birds on the island are a species of thrush (*Turdus Guianensis*?), a bunting (*Emberiza Brasiliensis*?), and the common moor-hen (*Fulica Chloropus*). These birds have spread over the whole island, and are found on the table-land as well as on the low ground. The *Fulica* conceals itself in the wood, where it is occasionally run down by the dogs."

Dugald Carmichael is my third eyewitness to see the Moorhens of Tristan da Cunha. It is remarkable he does not mention that these birds could not fly. Maybe he never noticed, because our Common Moorhens also have the habit, when disturbed, to run to safety, instead of flying. For Carmichael, they were just ordinary Moorhens (that he calls them *Fulica* instead of *Gallinula*, is of no importance).

Carmichael brought his collection to England, depositing his specimens in the London Museum. Unfortunately, this museum no longer exists. Many of his bird skins ended up in other museums. Most of his seabirds are now in the British Museum at Tring. The thrush and the bunting are in Berlin. The 'ordinary' Moorhens Carmichael collected on Tristan have vanished.

7. The London Museum

When Napoleon met his waterloo in Waterloo, his chariot was confiscated, shipped to England and donated to the British Prince-Regent, who either had no interest or needed money. He sold it to a certain Mr William Bullock for 3000 guineas. Bullock toured England and Scotland with it. This was an overwhelming success. More than 800,000 paying visitors came to see it, inside and outside. After the tour, Bullock deposited the chariot in his private museum, the London Museum.

The London Museum, also called the Egyptian Hall at Piccadilly, was a typical example of a nineteenth century museum of curiosities. It had two departments: the Roman Hall of Antiquities and Works of Art, and the London Museum of Natural History. The collections had been gathered over a period of more than thirty years, and contained objects from all over the world. There were many things collected by Captain Cook on his famous South Sea voyages. And there were Roman antiquities, such as two enormous mosaic bath floors from the palaces of Emperor Nero. There were marble tables with legs in the shape of golden dolphins, ivory ornaments from Africa and Asia, art objects from the Aztecs and Mayas from Mexico, a scale model of an elegant Chinese lady's leg with shoe, a garden chair made of a whale vertebra, and many more. Bullock also had a large collection of paintings, many depicting the heroic achievements of Napoleon, And, of course, the *piece de resistance* was Napoleon's chariot.

The natural history department contained valuable shells, geological and archeological objects, and hundreds of stuffed birds and mammals, often grouped in dramatic scenes, like the Bengal Tiger which was being strangled by a six metre Boa Constrictor (the Tiger came from India, the Boa from South America, but never mind). There were also some very rare specimens, such as the now extinct

Barbary Lion which used to live north of the Sahara.

Birds made up the greatest part of the collection. Bullock had all the parrots and parakeets that had been collected by Joseph Banks, during his voyage with Captain Cook (1768-1771). Many species in his collection had not yet been described. Bullock had also travelled and collected many birds, and even had new species named after him, like Bullocks Oriole in North America. His must have been one of the best and most complete collections of his time.

In 1819 Bullock decided he had enough. In a public auction sale, lasting 26 days, he sold everything. Unique collections were ripped apart and dispersed around the world. The auction sale drew a lot of interest. Buyers came from all over Europe. On the first day, many paintings and curiosities were sold. Also on the first day, Nero's bath floors were bought by an Italian gentleman for 249 and 357 pounds respectively. Where these mosaics ended up is unknown.

More than half of the sale was devoted to birds. Selling all Joseph Banks' parrots took a whole day. It is not surprising that curators of other European natural history museums were at the sale. For the British Museum there was Leach, for Leiden, Temminck, for Berlin, Lichtenstein.

The highlight of the auction came on the last day: Napoleon's chariot. The empty carriage was sold to a coach builder, who could convert it to something useful, for 168 pounds. Then followed the Emperor's bed, his famous overcoat, a silken undercoat, hats, caps, pistol holders, uniforms, his medals, and more. Mr Losky and Mr Lincoln both bought a tooth brush for three pounds. The razor went for eight pounds. Then there were soap boxes, mirrors, and of course the golden tongue scraper and ivory box with the sandalwood toothpicks. Thus, Napoleon's belongings were hopelessly dispersed, divided by more than 50 buyers, for a total of less than 1000 pounds.

Not all buyers were celebreties, representing other

museums. For instance, there were Mr. Winn and Mr. Ryall. Winn purchased a lot of articles from the South Sea, brought home by Captain Cook from Hawaii, Tahiti, and New Zealand: war clubs, feather hats, nose flutes, and the feather cloak, presented to Cook by the King of Hawaii. And a fly flap, the handle made out of the bone of a chief taken in battle. Winn also bought some birds, mostly pairs of European species, like Avocets, Stilts, Corncrakes, Red-throated Divers, Gadwalls, Skuas, and a pair of Moorhens.

Ryall bought various curiosa, such as a razorblade from Napoleon, a cap made of the skin of a Wild Boar, three stuffed boas, and various birds, with no clear plan. He obtained a Red Phalarope, a Purple Gallinule, a pair of Pied Flycatchers, an eagle, a Ringed Plover, a Turnstone, crows, a Red-necked Grebe, and one Moorhen.

What happened to Winn and Ryall, and the birds they bought, is unknown. The three Moorhens they purchased were the ones Dugald Carmichael had collected on Tristan da Cunha two years earlier.

8. Blenden Hall

Shipwrecks play an important role in the history of Tristan da Cunha. Survivers sometimes decided to stay and become part of the growing population on the island. Shipwrecks always meant a welcome supply of timber and other valuable materials. It is even said that girls would pray for a shipwreck, because it would enable them to get married (boys never pray for such a thing). The most famous and colourful shipwreck, no doubt, was the *Blenden Hall*, which was dashed to pieces on the rocks of Inaccessible Island in 1821.One of the passengers was Alexander Greig, the son of the captain. He kept a diary of the events, written in penguin blood. In July 1821 the Blenden Hall left England for Bombay with 24 passengers. At the last moment, Lieutenant Painter brought along his new wife. Soon it became obvious that she was not a real 'lady', but just a pilot's daughter, a disgrace for the better circles. Being a foot taller than her husband, she addressed him as 'little Painter', which gave rise to e great deal of hilarity. Greig writes:

"from the first moment after Mrs. Painter opened her mouth, I was satisfied that her claim to the title of *lady* had been established as recently as her marriage, and was sorry that the lieutenant had been caught by a *pretty face* only; but when I heard her address her husband in such a disrespectful and indecorous manner, I really pitied the man."

There was another not so ladylike lady on board, Mrs Lock, the wife of a high ranking military man in Bombay. She was an even bigger shock for the better circles than Mrs Painter because she turned out not to be white. She had booked through an agent, who had not mentioned this awful fact. She travelled with two little children, a 14 year old niece, Miss Morton, and a servant Peggy, about the same age as Miss Morton. Mrs Lock was ugly, corpulent, and spoke broken English. None of the gentlemen wished to sit next to her at dinner but she forced herself upon the doctor. She was seasick, and when the ship made a sudden lurch, she threw up all over him. For days she continued to apologise and begged him to sit with her again. Unsurprisingly, he refused. Mrs Lock and Mrs Painter found each other and started fighting. During the entire voyage they were heard to call each other 'whore' at the tops of their voices. At other times they behaved as though they were best friends. They both liked to command the lower ranking seamen. Another noteworthy passenger was quartermaster Hormby, with his 18 year old wife and baby. Hormby was a charming man, often amusing his fellow passengers with spooky stories.

Captain Greig thought it would be nice to visit Tristan. In thick fog the ship ran aground. The crew was taken by total surprise when the rocks loomed up from the mist, and there was no way they could avoid them. At that moment the wind completely died, the ship could no longer manoeuvre and the rudder became stuck in seaweed. A boat was lowered with a few strong men, who tried to pull the bow sideways to get clear of the rocks, but it was to no avail. The ship crashed on the reef, the rope snapped, and the boat disappeared in the mist. Although there was no wind, a strong swell caused heavy breakers, which started to take the ship apart. The captain gathered everybody on the forecastle, which was a good decision, because the stern was soon smashed to smithereens.

When the fog lifted it became clear that they were very close to the beach. A brave man swam to the shore with a rope, but he had to let go. Several men jumped into the sea to swim ashore. Two of them drowned. On the forecastle, they managed to construct a makeshift raft from wreckage, which unfortnunately was not big enough to accomodate everyone. Charming Hormby begged to reserve a place for his wife and baby, but at the last moment he jumped on the raft himself, and pushed off, leaving his wife and child behind. Furious men tried to throw him overboard, but he held on. They reached the shore in safety. Then an enormous wave threw the remains of the ship over the reef onto the beach. Miraculously everybody reached the shore alive. The first night was terrible. Everybody sprawled on the beach, totally exhausted, in heavy rain, without any shelter. Next morning a few men felt strong enough to get up and explore their surroundings.

Then the sea started to give back what it had taken. Among other things, a large cask with rum washed up. A swig of alcohol in a cold stomach works wonders. More people got up and started to collect wreckage, constructing shelters with the planks they found. But the rum also had a negative effect. Various crew members got drunk and began to fight each other and abuse the passengers. They had seen enough of the ladies and threatened to eat their children. Mrs Lock and Mrs Painter began their fighting again too. Mr and Mrs Hormby were no longer on speaking terms. Among other things, two enormous bales of red muslin washed up. This was a welcome gift, as many people had their clothes ripped to shreds. They made comfortable ponchos and everybody resembled Father Christmas.

To their horror, they saw how a sailor killed a bird and ate it raw. But everyone was hungry and soon followed his example. First they took the thrushes that were feeding on an Elephant Seal carcass. Soon they also started to kill penguins, which was not always easy. Many people tried to make a fire by rubbing sticks together without success. Then a chest with medical supplies washed up containing flint and steel. Finally, they managed to make a big fire and soon everybody started to roast whatever they could find. Whole seals were thrown into the fire, sometimes not even dead yet. Gradually, the kitchen became more refined. A buoy was cut in half to make two pans. The best meat came from the Elephant Seals, small pieces wrapped in cloth, with wild celery, cooked in sea water. The tongue and liver were especially valued. After a week some men managed to climb to the top of the island. It was then that they found out they were on Inaccessible, seeing Tristan on the horizon. They also found albatrosses, a welcome addition to the menu.

Captain Greig had arranged that the two ladies were housed as far apart as possible, but when a gale destroyed their tents, they managed to construct new ones close together again, so the arguing continued. The crew was split into two camps. Some stayed with the captain, and helped the passengers, but others had decided to no longer take orders, and in their separate camp they spent most of their time drunk. More casks with rum washed up, and after a few days, also boxes with wine. Hope flared when the boat was found which had been used in the attempt to pull the bow sideways. The men had said it was smashed to pieces when they reached the beach, but now it appeared they had just rowed ashore, and pulled the boat to safety. With this boat, they could go fishing, approach passing ships, or, on a calm day, even reach Tristan. But soon, when it was not pulled up far enough, a gale reduced it to firewood.

One of the ringleaders was Jack Fowler, who almost daily threatened to roast the Lock children. One day he shouted "ship ahoy!", and when everbody ran to the beach to look, he broke into the Painter residence to steal money and jewelry. He was accused, but he pointed to other possible culprits, and said that Painter had probably hidden the loot himself. Greig appointed a comittee to inspect all the tents. Jewelry and money were not retrieved but other interesting things were found. Mrs Painter had stolen a box with silken shirts from the doctor, and charming Hormby had a secret store of medical supplies, food, and booze. They saw no passing ships, so after four months it was decided that their only hope would be to reach Tristan themselves. Three competing groups started to build their own boat, stealing each others nails at night. The first party did not make it and drifted away. They were picked up by a British whaler, which did not take the trouble to sail to Inaccessible to look for the others. The second group managed to reach Tristan and the rescue of the rest could start. There were too many to be transported at once, so it took several trips to get everybody to Tristan, where they greatly outnumbered the resident population. At that time, there was still only the Glass family and a handful of bachelor friends.

On Tristan the fighting continued between the ladies and between the crew and passengers. Things escalated when William Glass asked six men to go fishing. They refused because they had sworn never to do anything again for the benefit of the passengers. In response Glass refused to supply the crew with any more potatoes and locked his storehouse. Some sailors, led by Jack Fowler, tried to break in but this was prevented by Greig and the men who had remained faithful to him. Jack was jailed and the others retreated in a distant camp. Finally, in January 1822, the *Nerina* and the *Susannah* were able to take everybody away. The islanders were immensely relieved. Two people stayed behind on Tristan: crew-member Stephen White and Peggy, Mrs Lock's maid. Stephen and Peggy had fallen in love, become engaged and wanted to start a family on Tristan.

Alexander Greig's book was published 25 years later. He used aliases for his characters, to prevent being sued by them. There are two other reports of the events. One was found in the pension in Cape Town, where the passengers had stayed. It ended up in an auction sale, and was published in the *Cape Monthly* in 1857. The author is unknown, but he gives the names of the other passengers:

"Mrs. Capt. Keys and two children, Lieut. and Mrs. Pepper, B.M., Quartermaster and Mrs. Gomsby and child, Miss Harris, Messrs. Law and Patch, assistant surgeons, Messrs. McTavish, Liddel, Lenon, Gibburn (cadet), Newman, B.M., Furlong, four mariners, and myself."

It is obvious that Mrs Key is Mrs Lock, Pepper is Painter, and Gomsby is Hormby. Like Greig, the author tells us how bad the crew behaved, especially after they found alcohol. He does not mention the arguing ladies. The third report was written by Pepper, AKA Painter. He was 27 years old, and his new wife, Ann, only 21. Pepper is even more careful with his words. In later Pepper generations several girls were named Nerina, after the ship that delivered them from Tristan da Cunha.

9. August Earle

Artist August Earle was born in London on June 1st 1793. Both his parents came from America. His father, James, was a painter, specialising in portraits. August started painting at a young age, and had his first exhibition when he was 13 years old. At the age of 21, he finished his education, and started to travel. Via the Meditteranean, North Africa and North America, he ended up in Brazil, where he lived for several years. On February 17th, 1824 he left Rio de Janeiro. He planned to go to Calcutta, where he hoped to find work with Lord Amherst, the governor-general of India. It would take a while to get there...

The ship that would take him first to Cape Town, the *Duke of Gloucester*, would touch Tristan da Cunha, on the way. Captain Amm (sometimes spelled Ham) was a friend of the Glass family, and he had mail for them. On March 6th the island was within sight, but due to a strong SE wind they could not approach. Then the clouds descended to sea level and they bobbed around in the fog until after four days, they saw the contours of Nightingale. They were swallowed by the fog again, until finally on the 18th they could once more see the island, twenty miles to the southwest. The wind was favourable so they approached the east coast, where they could distinguish the huts at Sandy Point. In beautiful weather and with a flat calm sea they sailed along the coast to the north, and soon they would be able to see the Settlement around the corner. But sudden strong squalls (williwaws or willies, as the islanders call them) from the mountain almost caused them to capsize, so they had to retreat from the island again. Then a gale sprung up and they spent the night near Inaccessible. All next day the weather was terrible with no sign of an island anywhere. On March 20th they saw land again. It appeared to be Inaccessible. They set course for Tristan, but it would still take another six days to finally reach the island. They landed on the 26th, twenty days after they had first sighted the island.

In all likelihood Captain Amm was a bit nervous, and he probably told everyone who went ashore to keep one eye on the ship at all times and to be prepared to board immediately if he hoisted the signal flag. Tristan produced potatoes of high quality, and Amm was willing to take a few tons to Cape Town to sell them on behalf of the islanders. Loading would take two days, enabling Earle to spend some time ashore to look around. He took his dog and his painting equipment. When the loading was done Earle went to the beach to board the ship, but to his astonishment he saw how the *Duke of Gloucester* moved away and disappeared. He waited for two days, in heavy gales, but when the weather cleared, there was no ship in sight. He realised that he was marooned on the island, and would have to wait for another ship to rescue him. He was not the only one left on Tristan. With him was sailor T. Cooch, who did not mind staying there. He knew his payment would accumulate to a nice sum when he eventually returned home. Earle also adjusted happily to his fate and started to paint, expecting to find passage on another ship soon. It turned out that this would take eight months.

Earle was impressed by the wild nature of the island, and admired the islanders, who were able to manoeuvre through the waves with their featherlight boats. He painted, and made day trips with the islanders. They went to Sandy Point to plant potatoes, visited remote beaches to slaughter Elephant Seals and collect penguin eggs, and climbed the mountain to hunt wild goats and pigs. There were numerous goats at higher elevations. When they separated one using a dog, they would kill it, carry it to the edge of the cliff, remove the entrails, stuff it with ferns, and throw the carcass over the cliffside, where it would find its own way down. On the mountain, they found not only goats, but also albatrosses. About the albatrosses Earle wrote:

"This bird is the largest of the aquatic tribe; and its plumage is of a most delicate white, excepting the back and the tips of the wings, which are grey: they lay but one egg, on the ground, where they form a kind of nest, by scraping the earth around it; after the young one is hatched, it has to remain a year before it can fly; it is entirely white, and covered with a woolly down, which is very beautiful. As we approached them, they clapped their beaks, with a very quick motion, which made a great noise. This, and throwing up the contents of the stomach, are the only means of offence and defence they seem to possess; the old ones, which are valuable on account of their feathers, my companions made dreadful havoc amongst, knocking on the head all they could come up with. These birds are very helpless on the land, the great lenghts of their wings precluding them from rising up in the air, unless they can get to a steep declivity. On the level ground they were completely at our mercy, but very little was shown them, and in a very short space of time, the plain was strewn with heir bodies, one blow on the head generally killing them instantly."

The human population on Tristan, when Earle arrived, consisted of six adults and a couple of children. (There had been more inhabitants. When the *Blenden Hall* floundered there were around a dozen men. Half of them sailed to Cape Town to do business for the island but never returned). The adults comprised four men and two women. Corporal Glass and his wife were the only ones remaining of the original settlers when the garrison left in 1818. Glass had been living there now for six years with his wife Maria Magdalena (called Mary), and they had produced a child almost annually. The second couple was Stephen White, with his young wife Peggy, who stayed when the other castaways from the *Blenden Hall* left with the *Nerina*. Peggy delivered a fine baby girl while Earle was there. Then there were two bachelors, John Taylor and Richard (Dick) Riley, who lived in the so-called Bachelors

House. Earle stayed with the Glass family, sailor Cooch lived with the bachelors. Being well educated, Earle was appointed schoolteacher and taught the children to read and write, something most of the adults were unable to do. And he was asked to hold a religious service every Sunday.

Earle's diary begins happily. He admires the nature, likes his friends, and enjoys the excursions around the island and on the mountain. But gradually, when the months drag by, his tone changes, and he apparently suffers from depressions, especially when a ship is seen at the horizon that does not stop. He no longer writes about this beautiful island, but this miserable island, instead. Like:

"Today I completed six months' miserable imprisonment on this wretched island, and have no more prospect of getting off than I had the first week that I came on shore. Instead of becoming reconciled to my situation, I think I am lately more and more wretched: every species of pastime or occupation I could think of or invent, I have exhausted. I sit for hours together watching the horizon, with the faint hope of catching sight of a vessel, and thinking of my friends in England. Previous to the return of Spring, my gun was a source of amusement, though my game, generally speaking, was no better than gulls and various kinds of aquatic birds: but now, even that employment is denied me. This being the breeding season, they strew themselves in all directions about the island; and as they place their nests in the most exposed situations, it totally destroys any pleasure I might have in the pursuit of them; for, however unaccountable, it is the fact, that the principal pleasure of shooting is the excitement, the uncertainty, and difficulty of following and bringing down your prize. Now, that I am so surrounded with birds, that I might easily take a waggon load with my hands, I do not feel the slightest inclination to touch any of them."

Finally, on November 29th, he was able to board the *Admiral Cockburn*, heading for Van Diemen's Land (Tasmania). Earle stayed in Australia for a couple of years. In 1827 he decided to paint the famous head hunters of New Zealand, joining an expedition of missionaries. He stayed with the Maoris for nine months. In 1828, he finally arrived in India, with a grand detour via various South Sea islands. Health problems forced him to return to England, where in 1830 he began writing about his travels. His book *Narrative of a nine months' residence in New Zealand, in 1827; together with a journal of a residence in Tristan d'Acunha, an island situated between South America and the Cape of Good Hope* was published in 1832, by which time its author was travelling again. In 1831 he received an offer to join Darwin and the *Beagle* as an artist, which would have made him very famous had he stayed on board. But recurring health problems forced him to leave the ship in Rio de Janeiro. He died in England in 1838.

Earle's *Narrative* raised a good deal of dust, because he openly criticised the missionaries and their attitude towards the Maoris. Many of his watercolours are now in the museum of the National Library in Canberra, Australia. These include the sketches he made on Tristan da Cunha. Several of these have been used for designing stamps of Tristan.

Earle was not an ornithologist. He recognised albatrosses and penguins, but aparently had no idea what an ordinary British Moorhen might look like. On one of his trips to the mountain his dog killed some big, fat albatrosses. Their meat tasted like the finest lamb and their fat was excellent for frying. The dog not only killed albatrosses but also Moorhens. The brown birds, which Earle thought were females, were in fact juveniles. In Moorhens, male and female look alike. Earle gives a perfect description:

"Besides our albatross, the dogs caught some small birds, about the size of our partridge, but their gait was something like that of the penguin. The male is of a glossy black, with a bright red, hard crest on the top of the head, The hen is brown. They stand erect, and have long yellow legs, with which they run very fast; their wings are small and useless for flying, but they are armed with sharp spurs for defence, and also, I imagine, for assisting them in climbing, as they are found generally among the rocks. The name they give this bird here, is simply "cock," its only note being a noise very much resembling the repetition of that word. Its flesh is plump, fat, and excellent eating."

10. Willem Luijke

People with a fascination for Tristan da Cunha are found all over the world. I even found one in my own home town, the retired Reverend Botte Kuiper. He acquired his infection when he found a travel report from a great grand uncle in a dusty box in the attic. In 1826 Willem Luijke was sent to the Moluccas by the Dutch Gospel Society, together with missionaries Gützlaff, Dommers, and Wienkotter (and spouse). Luijke stayed on Ambon until his death in 1866. He travelled with the ship *Helena Christina* with Captain Martense. On the way they made a stop at Tristan da Cunha. To my knowledge, the *Helena Christina* has not been mentioned by any other Tristan author, so it is my pleasure to tell the story.

The ship left The Netherlands on September 15th 1826. As early as the third day, they met their first storm. Luijke amused himself by studying the faces of his seasick fellow passengers. Looking at the sea, he admired the Greatness of God, but secretly he prayed to Him to keep the ship safe. There were tensions among the passengers between the Faithful and the Unfaithful, especially in relation to the heathen Neptune ceremony when crossing the equator.

On November 14th, the *Helena Christina* arrived at Tristan. They were approached by a boat with five men who came on board with a sac of potatoes, a quarter barrel milk, a whole pig and a half sheep. The captain went ashore with empty barrels to fetch fresh drinking water. This took several hours and enabled Luijke to briefly visit the island. It took them an hour to get safely ashore through the heavy surf and, when they landed, Luijke was soaked to the skin. He remarked that water was running through his trousers, but again, he praised the Lord, and exclaimed: "How great are your Works, my Lord! Thou created all this in Wisdom!" He hoped to have the opportunity to speak about Jesus on the island.

On the beach they were met by a man and several small

children. The man appeared to be English. They kissed the children, and each took one by the hand. When asked how many people were on the island, the man said there were eight adult men, two women, and eight children. Two of the men were married, and one was awaiting a wife who was coming from the Cape. Luijke tells us:

"One of the married men was as much as a headman, named William Glass, whose wife was born in the Cape, where her mother and her two eldest children were living. She spoke a little Dutch, but preferred English. The mother had served the well known reverend Vos. Because she could read a little Dutch, my booklets were a welcome gift. She had five children with her. The youngest two had been baptised by her husband, but she was very eager to have that done by a legal minister, but following my instructions I could not do that. The children were very sweet, with a brown skin and black eyes and hair. I walked the beach with the youngest, of a few months, on my arm, and the child smiled so sweet at me. Ik spoke to this woman and her husband, and the man who was awaiting his wife, about God, as the others were helping, filling our casks with water. They told me they came together every morning and evening reading an English prayer book and the bible. The man who was awaiting his wife showed me a quarto writing block in which he kept notes about what they had been reading. That morning, they had read Lucas 16, and I spoke in particular about Lucas 15 and Joh. 3. On Sundays they do not work, and read the bible. Last sunday, they had read the Book of Proverbs, Solomons 2, and Lucas 4. The woman's face clearly showed her interest in Religion.

William was here first, and is the head, and told us all this. He came here in 1816, when it was feared that Napoleon would escape from St Helena. Reinforcements came to the island, but were withdrawn. Given the choice, Glass and his wife chose to stay, and Eduard Teijlor. Some years earlier there were three Americans to catch seals, and take their skins. Two were dead, drowned when visiting another island; one was alive, named Kerrie, but he is now buried. In 1820 Stephen White came with his wife. They now have three children. His wife was from Bombay. They were stranded on another island, Inaccessible, where they were taken away by Glass. In the same year, Richard Raijlie joined them, after stranding with a ship. Of the rescued crew, he chose to stay. Shortly after that, John Taijlor, a British mariner, joined them at free will. In 1825, of the stranded Nassau, capt. Kas, two men stayed, Samuel Spnib, a young fellow of 17 years, and a certain Pieter Pieters from Altona. In May 1826, a Frenchman François Margriette came from Cape the Good Hope. Apart from him and Pieters, they were all English. How much of this is true, needs to be investigated."

Luijke also visited the graveyard, which as yet only had two graves. One was Currie (Kerrie in Luijke's story), the other one a young man who had not listened to good advice and had climbed the mountain on a Sunday, falling down a cliff on the way back. The text on his grave read:

"Sacred. To the memory of Ths. Brown: aged 19 a native of London who was killed by falling a depth of more than 100 yards March 2, 1823.
Take warning all who this may read,
On no day do an unlawful deed.
To the Sabbath pay all respect due,
Lest such a death befalleth you."

Luijke was very pleased to see how much veneration there was for God's Word, and the sacred Sunday. He asked Mrs Glass if she loved Jesus, and she said wholeheartedly "Yes!" But did Jesus love her too? She shrugged her shoulders and said she hoped so.

Finally, he interviewed Stephen White and his wife. They wanted to leave Tristan, but unfortunately, captain Martense had no space availabe on his ship. Mrs White could not read, but even though her English was poor, she could understand when the bible was read to her.

Then it was time to say goodbye. Before leaving they spent an hour fishing. Using just a line and a hook, they caught a total of 50, Rock Lobsters and two kinds of fish. Then they sailed away. In the Indian ocean they visited Saint Paul, a horseshoe-shaped remnant of a volcanic crater, offering a safe natural harbour. In fine weather, the passengers went ashore, and amused themselves shooting penguins. The ship arrived in Batavia on January 5th 1827.

11. Brides on the beach

Shortly after Luijke's visit, in December 1826, White and his family left Tristan. The reason they wanted to leave was that they did not get on at all with a certain Pert (also spelled Perl, Peart, or Pearl), who had recently arrived on the island. He was a criminal, who hoped to avoid trial by escaping to Tristan. Nobody liked him, but it seems that he particularly pestered the Whites. Luijke does not mention him, but if I try to account for eight men in his story, Pert must have been Samuel Spnib, using the name as an alias.

In November 1826, before Luijke visited, island friend Captain Amm brought two newcomers. One of them was a certain Taylor (not John Taylor, see below - this is confusing), who only stayed a short time, and about whom nobody writes. That must have been the Eduard Teijlor Luijke mentioned. The other was British, Thomas Swain. After an adventurous seaman's life, he was peacefully collecting seabirds' eggs at the Cape where Amm found him. Amm thought that living on Tristan would be something for him. He was right. Swain stayed there for the rest of his life, leaving many descendants. After Glass, Swain is the second surname still there today. Swain was a marine, who allegedly held Admiral Nelson in his arms when he died in the battle of Trafalgar. As a prisoner of war, he had served in the French army, and called himself Francis Marquet. He must have been the Frenchman Luijke mentioned. During Luijke's visit there were also John Taylor and Richard Riley, who were already there when Earle stayed on the island. John Taylor's name was actually Alexander Cotton, but he used Taylor as an alias. Finally, there was Peter Peterson, or Pederson, from Denmark, called Pieter Pieters from Altona by Luijke. When White and the other Taylor left, there were six men on the island, five of whom were bachelors.

When Amm brought Swain to Tristan, he also promised to try to find him a wife, so Swain was probably the man who, according to Luijke, was awaiting a wife. Amm intended to try to find wives for all the bachelors on Tristan, for twenty bushels of potatoes per woman. On March 9th 1827, he sent a request to the governor of St Helena to support the small community on Tristan. He did not ask for women, but for agricultural equipment. He mentioned that there was only one married couple on Tristan, but that the circumstances would be favourable for four or five more families. Amm offered to take interested families to Tristan. In the meantime he found five women who were willing to undertake this adventure. They boarded the *Duke of Gloucester* on March 17th. Amm notified the governor, who replied that he was worried about the welfare of the women. He would not give permission to sail if there was no guarantee that the ladies could find suitable employment as maids, as they were not married, and one of them even had children. Amm did not reply and sailed away.

On April 12th he deposited his peculiar cargo on the beach of Tristan da Cunha. There were four young mulatto girls, and a 'negress' of over thirty with four children (I know the words mulatto and negress are no longer politically correct, and we are not supposed to use them anymore, but I do so here simply because I quote them from old sources).

There are several different stories about what happened next on the beach. One story has it that Amm landed the women at daybreak and quickly sailed away, before the men discovered them, afraid that they would not be accepted and be sent back. Another story is that Swain exclaimed that he would accept the first woman that set foot onshore. And that happened to be the black woman with the children. Yet another story tells that the men were given the choice by seniority, not of age, but by the time they had already spent on the island. Thus, Alexander Cotton (AKA John Taylor) had the first choice, followed by Riley, Peterson and Perl, in that order. They all chose the young girls, so Swain was left with the black woman and her children. We don't know what really happened as the stories are conflicting.

Whatever the true story, all marriages turned out happy and very productive in terms of children. A census in 1851 revealed nine families, with 64 children. Of the five couples that were formed in 1827, three were still present: Swain, Cotton, and Riley. Peterson died in an accident in 1832, and his widow and children left the island. Perl and his wife left in 1837 which was a great relief for the others. There was still the ever-growing Glass family. The remaining five families in 1851 were all formed by marriages with the daughters of the first settlers, including three daughters of William and Maria Glass. Later, the Rileys also left.

The story of the brides on the beach has been told many, many times. Everyone writing about Tristan brings it up as it is such a good story. It is often told with a great deal of humour, as if the whole thing was just one big joke. But when you start to unravel their personal histories (see chapter 15), they come to life as real human beings, with feelings and ideas. Of all the dozens of people who ever came to Tristan, to settle for longer or shorter periods, these five women have been the bravest and the most adventurous by far.

12. The Stoltenhoff brothers

Frederick Stoltenhoff was born in 1847, Gustav in 1852. Their mother was English, their father German. Engelbert Stoltenhoff was a wool dyer, who had come to England to learn the newest techniques. There he met and married Margaret Pickersgill. They moved to Russia, where the Czar wanted to develop wool dying as an industry. Later he founded his own enterprise in Aken, Germany.

The two brothers were very different. Frederick wanted a good education and a steady job. Gustav, on the other hand, wanted to see the world. What they had in common was their hopeless love for Hannchen, daughter of textile merchant Emanuel, for whom Frederick worked as a clerk. Hannchen liked them both very much, but did not want a relationship with either of them. Frederick and Gustav both got married eventually, but secretly, deep in their hearts, they never could forget Hannchen. She never married.

When he was seventeen Gustav went to sea, on board the *Wilhelmina*. He had a tough time, survived his first gales, and was addicted to sailing forever. Frederick had to serve in the war between Prussia and France in 1870. On the first day at the frontline he saved the life of his sergeant which earned him a promotion. After the war, in 1871, he was honourably discharged.

On August 4th 1870, while his brother was at the front, Gustav sailed on the *Beacon Light* from Greenock, Scotland, with a cargo of coal for Rangoon. His fellow sailors told Gustav about whaling, which was very profitable. Some whalers earned a fortune in just two years. Even better was sealing, Scotchman McTavish told him. Sealing had the advantage that you did not need a ship. You just had to be dropped off and picked up again in a place where seals were abundant. Selling their beautiful skins could make you very, very rich in a short time.

South of Saint Helena the ship caught fire. The crew tried to smother the fire by closing the hatches and seal the seams with rope and tar, but to no avail. The redhot deck bulged, like a volcano, and an explosion followed. One sailor was blinded, and others burned, but nobody died. They all got in the lifeboats and watched from a distance as the ship went under in sizzling steam. The nearest inhabited place was Tristan da Cunha, 600 miles further south. They sighted the island on a calm Sunday and soon saw the islanders approaching them in their boat. They were hosted by several families. Gustav helped his host work the Potato Patches and patched up a damaged roof. One morning he woke up to find that his host and a few others had left for Inaccessible for a couple of days. He very much regretted they had not taken him with them. He was told that sealing was very profitable there too. Last year 1700 Fur Seal skins had been taken on Inaccessible, worth a fortune! Gustav found that the stories McTavish had told him were true. Inaccessible was paradise on earth.

Gustav eventually returned to Aken where his parents lived. Coincidentally, Frederick returned home from the war on the same day. They had a lot to tell each other. Pa Stoltenhoff told them that Hannchen had received a marriage proposal from a nobleman. This was a huge blow for both of them so they decided to leave. Gustav had convinced Frederick that their future lay on Inaccessible and that was where they should head. Their family was sceptical at first, but eventually helped them with their preparations. Uncle Carl gave them a hunting rifle and Hannchen's father donated a large quantity of seeds, so they could grow their own cabbage, beans, peas, lettuce, tomatoes, and carrots. They bought a wheelbarrow, spades, buckets, axes, a pick-axe, and carpenters' tools. Father Stoltenhoff gave them a large quantity of flour, dry bicuits, coffee, tea, sugar, pepper, tabacco, and fourteen bottles of Dutch genever. Hannchen's father wrote a letter of introduction to a shipping agent in England whose ships regularly sailed to Saint Helena.

In August 1871 they travelled via Rotterdam to England and then to Saint Helena, where they bought more building materials and a small boat. On November 6th they boarded the *Java* from New Bedford. The captain knew Inaccessible very well and he very much liked the brothers' plan. He said that he wished he could stay with them. They would not only find large numbers of Fur Seals and Elephant Seals, but also plenty of birds and their eggs to fill their cooking pots, as well as fat pigs. On the 27th of November the brothers, with all their luggage, were deposited between the penguins on Salt Beach, a narrow strip of land on the eastern shore of Inaccessible, at the foot of an incredibly steep cliff. As a farewell present the captain gave them a sack of potatoes and a bitch with two puppies to keep them company.

Thus started the adventure of the brothers Stoltenhoff on Inaccessible. They stayed there for two years and experienced nothing but problems and hardship. There were hardly any seals, and when their provisions ran out, and the birds and their young had left the island, they had a great deal of trouble filling their cooking pot. By climbing a narrow chimney in the face of the cliff, pulling themselves up on clumps of tussock grass, they were able to reach the plateau to hunt goats and pigs, which was not easy. The pigs' meat was virtually inedible, with a strong fishy, oily taste, because the animals fed on seabirds. The goat meat they found extremely delicious. Then disaster struck. The tussock on their climbing route caught fire and they could no longer climb up. With their boat they could reach the western side of the island, at Blenden Hall Beach, where it was much easier to get to the plateau, but that also became impossible when their boat was lost in a gale. After that one of them managed to swim to the other side on an exceptionally calm day where he ascended the plateau to hunt. He threw the animals down the cliff to his brother.

One day people from Tristan arrived and, much to their surprise, found the two men. They invited them to join them on Tristan, but when the boys declined the offer the previously friendly islanders became unfriendly. On another visit they killed the last Fur Seals and Elephant Seals, and shot all the goats on the plateau, thus depriving the brothers of their livelihood. During the winter they almost starved, but eventually spring came, bringing birds and eggs and a few seals arrived on the beach. Their optimism returned and when the *Java* came back to collect them they chose to stay and try another season. But their second winter was even worse than the first. In two years they had only gained one and a half barrels of oil and two long beards.

In October 1873 they gave up. The Expedition ship *Challenger* visited, bringing scientists to Inaccessible. The offer to board the ship was too good and there ended the Stoltenhof adventure. They left the ship in Cape Town, where they grew old. They both married, but still dreamed of Hannchen. Ironically, Hannchen also moved to Cape Town, where she remained a spinster for the rest of her life. Her nephew Eric Rosenthal described the story of the two brothers in his book *Shelter from the Spray*.

In 2005 I was approached by Hans Fredy Stoltenhoff, a successful Dutch businessman with offices in Belgium and the Canary Islands, and distantly related to Gustav and Federick. He was writing a book on how to become a successful entrepeneur, with his memoirs, including a chapter on Tristan da Cunha and the Stoltenhoff adventure. He asked me to comment on the manuscript. He thought that Inaccessible had been renamed Stoltenhoff Island, in honour of the two brothers. I had to disappoint him, and tell him that the name was only given to a small, at that time still un-named, rocky islet next to Nightingale, by Captain Nares of the *Challenger*. I was able to send him some pictures of Tristan, Inaccessible, and Stoltenhoff Island.

13. The Challenger expedition

The British Navy not only engages in warfare, but also has a long tradition of scientific expeditions. Scientists with adequate funding could use naval vessels in peace time. The *Challenger* expedition was the initiative of professor Wyville Thompson of the University of Edinburgh, and the Royal Society in London. The government provided the ship and the funding. Wyville Thompson was the expedition leader.

The *Challenger* expedition made history. From late 1872 to early 1876, in more than 1000 days, the ship sailed almost 70,000 miles, from England to the Caribean, from New Foundland to Cape Town, through the southern Indian Ocean, around Indonesia, Australia and New Zealand, through the northern and southern Pacific, and Strait Magellan in the far south of Patagonia. At 362 locations oceanographic, meteorologic and marine biologic data were obtained. The deep sea bottom was dragged and yielded many new life forms. Soil and seawater samples were collected, and thousands of creatures disappeared into jugs containing alcohol or formalin. It took scientists twenty years to analyse the data. The results were published in fifty voluminous volumes, folio format. Volume 1 was published in 1885, volume 50 in 1895. The expedition was also one of the first to take a photographer on board, giving rise to unique pictures of people in distant islands and kingdoms, such as the Japanese Ainoes in full pontificals, the royalties of Tahiti, and the now extinct Indians of Tierra del Fuego. But there are also hundreds of minute drawings, in black and white and in color, depicting newly discovered marine organisms and microscopic algae, and also artists' impressions of islands, icebergs, people, and thrilling events.

There were several biologists on board who were mostly marine biologists. Henry Nottidge Moseley had no such specialty. He had the more general title of 'naturalist', observing everything. He was bored by the long sea days and the endless dragging of the sea floor. He enjoyed his days ashore, especially on exciting islands. He found something new everywhere. Even under a pile of discarded banana leaves in a dull city, he would look for new species of flatworms.

In October 1873 the *Challenger* arrived at Tristan. In the Settlement they counted fifteen houses and 84 people. Moseley went ashore to collect plants. As the weather was unreliable, he was not allowed to go any further than half an hour from the landing place. With a small boy as a guide, he went a little way up the steep slope, where he was suddenly surprised by a strong squall, chilling him to the bone. The boy just sat down, folded his arms around his knees, and waited stoically for the squall to pass. Moseley was deeply impressed by the natural way the boy handled the weather.

After Tristan, they went to Inaccessible, where they arrived at night after dark. All night they heard the terrible noise made by the penguins. The next morning they went ashore and fought their way through the penguin colony. They also found the Stoltenhoff brothers, who were all too happy to join them, and leave the island. Mosely thought had he discovered a new species of very small Dolphins. He writes:

"As we approached the shore, I was astonished at seeing a shoal of what looked like extremely active very small porpoises or dolphins. I could not imagine what the things could be, unless they were indeed some most marvellously small Cetaceans; they showed black above and white beneath, and came along in a shoal of fifty or more from seawards towards the shore at a rapid pace, by a series of successive leaps out of the water, and splashes into it again, describing short curves in the air, taking headers out of the water and headers into it again; splash, splash, went this marvellous shoal af animals, till they went splash through the surf on to the black stony beach, and there struggled and jumped up amongst the boulders and revealed themselves as wet and dripping penguins, for such they were."

Moseley noted that most of the penguins landed at the same spot, where a wide alley led into the colony. From the main street, side alleys would branch off, into the tussock. Moseley writes:

"With one of the Germans as a guide, I entered the main street. As soon as we were in it, the grass being above our heads, one was as if in a maze, and could not see in the least where one was going to. Various lateral streets lead off on each side of the main road, and are often at the mouth as big as it, moreover, the road sometimes divides for a little and joins again: hence it is the easiest thing in the world to lose one's way, and one is quite certain to do so when inexperienced in penguin rookeries. The German, however, who was our guide on our first visit, accustomed to pass through the place constantly for two years, was perfectly well at home in the rookery and knew every street and turning.

It is impossible to conceive the discomfort of making one's way through a big rookery, haphazard, or "accross country" as one may say. I crossed the large one here twice afterwards with the seamen carrying my basket and vasculum, and afterward went through a larger colony still, at Nightingales Island.

You plunge into one of the lanes in the tall grass which at once shuts out the surroundings from your view. You tread on a slimy black damp soil composed of the birds' dung. The stench is overpowering, the yelling of the birds perfectly terrifying; I can call it nothing else. You lose your path, or perhaps are bent from the first in making direct for some spot on the other side of the rookery.

In the path only a few droves of penguins, on their way to and from the water, are encountered, and these stampede out of your way into the side alleys. Now you are, the instant you leave the road, on the actual breeding ground. The nests are placed so thickly that you cannot help treading on eggs and young birds at almost every step.

A parent bird sits on each nest, with its sharp beak erect and open ready to bite, yelling savagely "caa, caa, urr, urr," its red eye gleaming and its plumes at half-cock, and quivering with rage. No sooner are your legs within reach than they are furiously bitten, often by two or three birds at once: that is, if you have not got on strong leather gaiters, as on the first occasion of visiting a rookery you probably have not.

At first you try to avoid the nests, but soon find that impossible; then maddened almost, by the pain, stench and noise, you have recourse to brutality. Thump, thump, goes your stick, and at each blow down goes a bird. Thud, thud, you hear from the men behind as they kick the birds right and left off the nests, and so you go on for a bit, thump and smash, whack, thud, "caa, caa, urr, urr," and the path behind you is strewed with the dead and dying and bleeding."

Mosely adds, that it is horribly cruel to kill whole familie of innocent birds, but that it is absolutely necessary to reach other areas to collect plants, illustrating that he was more a botanist than an ornithologist. It is quite ironic that the Northern Rockhopper, which he treated with so little respect, has been named after him *Eudyptes moseleyi*.

The Challenger expedition failed to see Moorhens, which apparently had already become very rare at the time of their visit. Expedition leader Wyville Thompson writes about them:

"...; and a singular bird called by the settlers "island hen," which was at one time very common, but which is now almost extinct. This is a water-hen, *Gallinula nesiotis*

(Sclater, Proc. Zool. Soc., 1861), very nearly allied to our common English moor-hen (*Gallinula chloropus*), which it resembles closely in general appearance and coloring, with, however, several satisfactory specific differences. The wings of the Tristan species are much shorter, and the primay feathers, and indeed all the feathers of the wing, are so short and soft as to be useless for the purpose of flight. The breast-bone is short and weak, and the crest low, while, on the other hand, the pelvis and the bones of the lower extremity are large and powerful, and the muscles attached to them strong and full. The island hen runs with great rapidity; it is an inquisitive creature, and comes out of its cover in the long grass when it hears a noise. It is excellent eating, a good quality which has led to its extermination."

The Germans reported that there was a peculiar Island cock living on Inaccessible, which the biologists failed to find. The brothers gave a good desciption, which was quoted by Wyville Thompson:

"Inaccessible, like Tristan, has its island hen, and it is one of my few regrets that we found it impossible to get a specimen of it. It is probably a *Gallinula*, but it is certainly a different species from the Tristan bird. It is only about a fourth the size, and it seems to be markedly different in appearance. The Stoltenhoffs were very familiar with it, and described it as being exactly like a black chicken two days old, the legs and beak black, the beak long and slender, the head small, the wings short and soft and useless for flight. It is common on the plateau, and runs like a partridge among the long grass and ferns, feeding upon insects and seeds. An island hen is also found on Gough Island; but the sealers think it is the same as the Tristan species."

Moseley was sceptical about the Island Hens of Inaccessible. He writes:

"We were in search of another land bird, a kind of Water-hen (*Gallinula nesiotis*), which is found on the higher plateau at Tristan, and is described by the inhabitants as scarcely able to fly. We could not meet with a specimen. Only very few inhabit the low land under the cliffs, and we were not able to land at the only place from which the higher main plateau of the island is to be reached.

The Germans said that the Inaccessible Island bird is much smaller than *G. nesiotis*, and differs from it in having finer legs and a longer beak. This is, however, hardly probable, since the Tristan species occurs at Gough Island."

Frankly, I don't understand Moseley's reasoning that Moorhens at Gough would make the occurrence of a different species at Inaccesible improbable. Anyway, he was wrong, and the little Cocks of Inaccessible were left to run around through the grass undescribed, for another fifty years. It is also noteworthy that they all knew about the Moorhens on Gough, almost twenty years before they were 'officially discovered'. It is also evident that the Island Cocks at Tristan were not extinct yet. They were just out of reach of the expedition. Yet, because the expedition failed to find them, and there have been no records by outsiders since, 1873 has gone into history as the year of extinction. The real extinction must have occurred later than that.

14. Ancestral mothers and fathers

There is only a handful of ancestral mothers of Tristan da Cunha. There only were seven. These were Maria Magdalena Leenders, Sarah Jacobs, Sarah's daughter Mary Jacobs, Maria Williams, Susanna Philips, Agnes Smith, and her sister Elizabeth Smith. Maria Magdalena (later called Mary) was the first, joining William Glass with the garrison in 1816. Maria has always been described as a coloured Cape Creole, and almost everybody writing about Tristan, reiterated that. Only Peter Munch, in his book *Crisis in Utopia*, has his doubts. He states that anyone born in the Cape could be called a Creole, regardless of skin colour. He suggests that Maria, with Dutch parents, could actually have been white. Amazingly, nobody ever mentioned her skin colour, but Willem Luijke found her children brown-skinned, black-eyed, and black-haired. Finally, Maria's African ancestry has been proven by mitochondrial DNA research (Soodyall *et al.* 1997).This has some importance, because nobody ever wondered whether she might have had her own motives to start a new life away from the Cape. Some authors think that her husband William wanted to get away because of racist remarks about his marriage. I don't know if this is true, but it is interesting to think that they may both have had their reasons to leave the Cape, and start a new life in a neutral place, where they could set their own rules and moral standards. Thus, Tristan may have become not only the first communist 'country', but also the first free of racism.

Born on January 1st 1801, she was incredibly young, only thirteen years old, when she married William Glass in July 1814. Her first child, William jr., was born in 1816, when she was fifteen. Nancy Hosegood, in her book *The Glass Island*, tells us how a gallant William Glass rescued her after she had been savagely beaten up by her wicked master, who she had to serve when she was only twelve years old after her mother died. I am afraid that Hosegood, as well as using solid sources, also relied somewhat on her romantic fantasy throughout the book. Maria told Willem Luijke that her mother was still living in Cape Town with her two eldest children, where she served as a servant of the wife of the 'wellknown' Reverend Vos. The fact that she was a servant suggests that her husband, Maria's father, the Dutchman Leenders, was no longer around.

Only two of the five Saint Helena brides left descendants on Tristan who are there to this day. The other three left successively with their families. The widow Sarah Jacobs, described as a Negress, and Maria Williams, described as a Mulatto, were originally said to be sisters. Later, based on their colour difference and Sarah's marriage certificate, they were declared half sisters. Their black mother, known as Delilah or Dilly Williams, probably first had a black husband, Mr Bowers, producing Sarah about 1795, and then a white one, Samuel Williams, producing Maria, born in 1810. Or perhaps Maria had a different father and took her mother's surname.

However, a study of mitochondrial DNA, which relates only to female lineages, has shown that in the founding mothers there was only one pair of sisters, instead of the two mentioned in all books (Soodyall *et al*. 1997). The Smith girls from Ireland, who arrived in 1908 (see below) were real sisters, but Sarah and Maria did not have the same mother. If they also had different fathers, as many believe, they were not sisters at all. Perhaps they used the term 'sister' in a non-European sense. Another alternative, if they both considered Delilah to be their mother, is that one of them may have been adopted.

Maybe Sarah Jacobs also had two husbands or partners. She may have been allied to a Mr Fisher, before she met Jacobs. When her eldest daughter Mary married Peter Green (see below), the name on her marriage certicate was Fisher, not Jacobs.

Sarah Jacobs married Thomas Swain, Maria Williams married Alexander Cotton AKA Taylor. Given the year the Saint Helena women came to Tristan, they, or certainly their mothers, must have known a past in slavery. Sarah's mother, at least, was listed as 'slave to D. Kay Snr'. Although slavery was already in the process of being abolished, Saint Helena must have been, like the Cape, a pretty rough racist environment, with very few prospects for coloured girls, especially unmarried mothers. This may have played a significant role in their choice to abandon their perhaps hated home island to start a new life in the arms of total strangers on a remote island in the South Atlantic. Of Sarah's four children, only her eldest daughter Mary, who married Pieter Groen (Peter Green), left descendants on Tristan. Two other daughters married men on Tristan but left the island with their families.

Actually, the story of the four children may be different. Later, a mysterious fifth child turned up. Fanny was the daughter of another one of the four Mulatto women, Sarah Bassett Knipe. Fanny married the Dane Peter Miller, who came to the island with the shipwreck of the *Emily*, together with Pieter Groen. The Millers left, and Fanny, like her mother, left no descendants on the island. Why was there never a fifth child mentioned by anyone? I think simply because there never were five, and Sarah Jacobs had three children of her own instead of four. I found this confirmed in a biography of Sarah on Wikitree.com. She came to Tristan with her three children, Mary, Catherine, and William. We have dealt with the daughters already. William simply disappeared. Why had nobody seen this? Perhaps, on arrival, the four children stuck together just because they were children, always seeking each others company. Or could it be that Sarah Jacobs temporarily adopted Fanny, perhaps to give the other Sarah a better chance to be a desirable bride? An indication may be that Jan Brander tells us that the Dane Miller married another daughter of 'the Negress'.

One of the sons of Sarah and Thomas Swain, Samuel, followed in his fathers footsteps and found himself a coloured woman from Saint Helena, Susanna Philips. She arrived in 1862. I know nothing of her background.

Finally, the last two of the ancestral mothers came to Tristan in 1908, the sisters Elizabeth and Agnes Smith, daughters of Ellen Gay, a Catholic from Mullingar, Co. Westmeath, Ireland, who married William Smith, a protestant British soldier, in 1875 in Mullingar Protestant Church. Elizabeth was born in Kilkenny in 1876, Agnes was born in 1887. The family moved from Kilkenny to Westmeath, or perhaps Meath, in Ireland, to England and then to Cape Town. The girls met the brothers Robert and Joseph Glass, both originally from Tristan. Elizabeth (Lizzie) married Robert (Bob) in 1895, Agnes (Aggie) married Joseph (Joe) around 1900. They were accompanied by a third sister, Annie, and her husband James Hagan, also a Tristanian, and all their children. The three families totalled sixteen individuals, arriving on Tristan with the *Greyhound* on March 26th, 1908. Annie and James did not stay long and returned to Cape Town. It was Agnes in particular, who promoted Roman Catholicism in Tristan.

Potentially, there is a new ancestral mother. In 2017, Kelly Burns, daughter of the longest ever serving administrator on Tristan, Sean Burns, married Shane Green. Whether she will become an ancestral mother depends on whether her children stay on the island and marry into the population, or leave the island before that can happen. Time will tell.

Given that the group of seven ancestral mothers contained one pairs of sisters, and one mother-daughter pair, we can reduce them to no more than five. The ancestral fathers are almost twice as many. There are eight in all. The first, of course, was William Glass. He told a lot of stories about himself to August Earle, during their endless evenings by the fire. William was born on 11 May 1786 in Kelso, Scotland. His parents were David and Janet Glassgow, not Glass. In 1804, allegedly after having been crossed in love, he enlisted in the army, lying about himself. He said he was 16 instead of 18, and his name was Glass instead of

Glassgow. As a soldier, the only war action he saw was an unsuccessful campaign in Germany. He was soon sent to the Cape, which was still a Dutch colony, under British occupation. The colony was ceded to Britain in 1814, a turning point in Tristan history, because it was shortly after that that the decision was taken to send the famous garrison to the island. William was with the artillery drivers, climbing to the rank of Corporal. He often entertained Earle about his colourful adventures in South Africa, including hunting parties and a terrible encounter with a huge lion, which he called tiger. As a Corporal, he led a group of Hottentot artillery drivers. He was impressed by their skills as excellent horsemen and fearless drivers. They would gallop along precipices that would make a white man tremble. He also praised their good humour and the only drawback was their tendency to drunkenness.

When William married Maria he was 28 years old, more than twice her age. When Earle asked him about his motives for staying on Tristan he did not mention racial problems in the Cape. All he said was:

"Why, you know sir, what could I possibly do, when I reached my own country, after being disbanded? I have no trade, and am now too old to learn one. I have a young wife, and a chance of a numerous family; what could I do better for them than remain?"

It is interesting to note that many sources tell us that The Glass family came to Tristan with two little children. This is incorrect. They arrived with one child only, their son William junior. However, a second child was on the way, who became the first child ever to be born on Tristan, the girl Mary Ann.

In spite of the egalitarian 'constitution', Glass became the natural leader of the community, always respectfully addressed as 'Governor'. In a later, revised edition of the constitution this position was even made official. Glass ruled as a true patriarch, and in later life his Christmas dinners were attended by over thirty children and grandchildren. William died of cancer on November 24th 1853. Three years later the entire Glass family moved to the USA, where already quite a few of the children were making a living in the whaling industry. For a while, the name Glass no longer existed on Tristan. Maria died on the 10th of October 1858, only two years after she arrived in America.

Of the second ancestral father, Alexander Cotton, AKA John Taylor, we know considerably less. He was born in Hull, in 1788 or 1789 (1802 is also given as his birth year), and had a successful career in the navy. He also served as one of Napoleon's guards at Saint Helena for three years. Again it was Earle, at the Glass fireplace, who heard his stories. During the occupation by the garrison, Taylor and a comrade belonged to a schooner, tending to the admiral. In this capacity they came to Tristan a few times. On one occasion, when they met Corporal Glass after the garrison had left, they came up with a great plan. They would go home, and, after being paid off, they would purchase things that would be useful to the farm and return to Tristan. They were paid off, but the temptations on shore were too strong to resist. When all the money was gone, they went to the Admiralty, demanding a pension, based on their long and faithful service. They would willingly waive that right if they were given a passage to Tristan. Their request was granted and they were put on board the *Satellite*, bound for India, arriving at Tristan in June 1821. Taylor's companion did not stay long. During Earle's visit, Taylor was one of the two bachelors. In 1827 he married the Saint Helena girl Maria Williams, who was then 17 years old. They produced twelve children; seven daughters and five sons. They lost two little boys who had accidents, and two grown-up sons in the life-boat disaster. Only his eldest daughter Martha, married to William Green, produced descendants. So after he died, around 1865, the name Cotton disappeared but the genes are still there.

The third ancestral father was Thomas Swain AKA Francis Marquet, who was taken prisoner of war by the French, and was forced to fight in the French army against his own people. When he was taken as a French prisoner of war by the English he never dared to reveal his true identity. He remained in prison for nine years, pretending he was French. When the war was over he was released and moved to the Cape, where captain Amm found him. In 1827, as we have already seen, he married Sarah Jacobs from Saint Helena. Thomas and Sarah had themselves baptised in 1851, and in the same year had their marriage legalised in Church, 24 years after they were unofficially united by Governor Glass. Apart from Sarah's own three children they produced another ten.

Thomas Hill Swain was born in Hastings. His birth date is a bit of a mystery. When he died in 1862, the headstone of his grave read that he was 102 years old, which would place his birth in 1760. But according to Brander he died at the age of 87, which would make it 1775 instead. When captain Amm found him in 1826, he was either 55, or 66 years old, which would make his birth year 1771 or 1760. As usual, I leave the mystery to the 'true' historians. His widow Sarah lived for another 31 years, passing away in 1893.

About ancestral father number four, the Dutchman Pieter Groen, we know much more, thanks to Jan Brander and other Dutch historians. Pieter was born in 1808, in Katwijk aan Zee. He was the second child and eldest son of Willem Pieterszoon Groen and Jaapje Jacobsdochter Schaap. Father William owned a bar in Katwijk. People in Katwijk mainly survived on herring fishing, and from a young age Pieter had to do his share in the fishing industry to provide food for the growing family. As a result, he was not able to finish school. But he was intelligent and taught himself a lot. In 1828, at twenty years old, he decided that his future would not be in herrings, but that he wanted to sail the world's oceans. He went to Rotterdam and became a sailor.

In 1833 there was a cholera outbreak in Katwijk, which killed Pieters parents, and his little brother Cornelis, who was only three years old. Of the surviving children, the younger ones went into an orphanage and the older ones had to take care of themselves. Pieter never returned to Katwijk. His career as a seaman did not last long, ending on the reefs of Tristan in 1836, as we know. He changed his name to Peter Green and married Mary Jacobs, producing a large family. When Glass died in 1853, Alexander Cotton, the 'next in command' as Earle put it, would have been the logical successor as headman, but in a sort of natural way, this role fell to Peter Green, who was much better educated and better suited to communicating with the captains of visiting ships. Green maintained his position as unofficial headman until his death in 1902.

In 1889, Tristan was visited by the English Poet George Newman. Green and Newman remained pen pals forever. Newman took letters from Peter to Katwijk, where he found his little sister Pietertje, 17 years his junior. There were also two surviving brothers. As a curiosity, Newman wrote and composed a 'national anthem' for Tristan da Cunha, but I am not sure whether the people ever adopted it or even know about it (see next chapter).

In 1973, the ties between Tristan and Katwijk were tightened. Tristan Administrator Fleming visited Katwijk in 1974, and in 1979 a secondary school was renamed Pieter Groen College. Like many other fishermen's villages Katwijk is a real bible belt town. Skippers refuse to learn to swim, believing that when God wants you to drown you drown. Some fundamentalist town council members were against the name change because they knew Pieter Groen was a communist. One of the pupils was Sandra van Duyvenboden (in later, married life Kornet-van Duyvenboden). When her history teacher explained the origin of the new name of the school she was hooked for life. She investigated the history of Groen/Green, and visited Tristan with her husband in 2001 to conduct her research, resulting in her book *The quest for*

Peter Green.

About ancestral fathers number five and six, the American whalers Thomas Rogers and Andrew Hagan, I know very little. There must be American historians, specialised in the whaling history of New England, who can tell more about them. Thomas (Jock) Rogers came to the island in 1836, a few months after the *Emily* was shipwrecked. He married Jane (Jenny) Glass, one of the daughters of William Glass. Two years later a ship called with a very sick man on board, Charles Taylor, who seemed close to death. The patient was taken ashore and Rogers took his place on the ship, leaving his wife with a little child. He promised to be away for one voyage only but never returned. Charles Taylor fully recovered and married another daughter of Glass. Deserted Jenny waited for years in vain. She lived a rather isolated life, in a poor shack away from the rest of the houses, on the other side of Hottentot Gulch, where she had a second child. It must have been a relief for her when she could leave, with her two children, joining the Glass exodus of 1856. Her son Joshua returned to Tristan in 1859, married Sarah Swain, and became the progenitor of today's Rogers clan.

Andrew Hagan, originally from Ireland, was not just a whaler, but a real captain. He had some unlucky voyages, hardly catching anything. When he reached Tristan in 1849, returning from South Georgia with an empty ship again, he decided to quit. The story has it that he hid in the bushes, and the mate gave up looking for him and left. Hagan married Selina Glass, whom he had met a few times before on previous visits. Selina was a good catch. When William Glass revised the 'constitution', he not only made himself headman, but also declared that all land would be the sole property of the first batch of settlers and their heirs. With others leaving or dying, this meant in practice that most of it would be his. The other islanders more or less ignored this. Everybody used the pastures without reference to property rights. But Andrew Hagan looked at it differently. After the 1856 Glass exodus, his wife Selina was the only Glass left on the island, and when she died, he considered himself the sole heir of everything, making him the richest man on Tristan. He built up a large herd, comprising half the cattle on the island, occupying most of the available pasture, much to the dislike of the others. He was nicknamed 'Cattle King of Tristan da Cunha'. Hagan married again, to Eliza Swain, daughter of Thomas Swain senior and Sarah Jacobs. His popularity waned and he was totally ignored by everybody. He committed suicide in 1898, cutting both his wrists with a razor and bleeding to death on the beach. He was 82 years old.

Finally, we come to ancestral fathers seven and eight, the Italians Gaetano Lavarello and Andrea Repetto, who stayed on the island after the shipwreck of the *Italia* in 1892. On board was also Agostino Lavarello, a cousin of Gaetano. During his short stay on Tristan, Agostino fell in love with Mary Green, but things did not work out and he went back to Italy. Mary wrote letters to him but, when he was at sea, these would fall into the hands of his mother who hid them, afraid he might emigrate to Tristan. After she died in 1915 he found them and this inspired him to write a book about the shipwreck, *I naufraghi di Tristan da Cunha*, which was published in 1936. In the last chapter, he declared his love for Mary. Had he followed his heart sooner and stayed on Tristan like his cousin, we would have seen two Lavarello's as ancestral fathers.

Andrea Repetto was born in Camogli on January 13th, 1867. He was a petty officer on the *Italia*, and before that served in the Italian navy. After being cast away on Tristan, Andrea married Frances Green, granddaughter of Peter Green and Mary Jacobs. They had seven sons, including Willie Repetto, who became headman of Tristan da Cunha after Peter Green and his father died. Andrea died on the 9th of February 1911, after a severe attack of asthma. He was only 44 years old. His widow, Frances Repetto, quite a character, became 'head woman', the uncrowned queen of Tristan da Cunha. Gaetano Lavarello was also born in Camogli, December 9th, 1867. He married Jane Glass, with whom he had six sons. He died August 9th, 1952, almost 85 years old.

As in Katwijk aan Zee, the people of Camogli know about Tristan da Cunha. Camogli is a picturesque Mediterranean fishing village and holiday resort on the Ligurian Riviera, not far east of the city of Genoa. There is a nautical museum with a large show case dedicated to Tristan da Cunha.

In the early 1990s surname number eight arrived at Tristan. Pat Patterson had served several years on Tristan as postmaster and radio man. He married an Island girl, Susan Green, with whom he left the island. After his retirement he got permission to establish himself and his family on Tristan. He was there when I visited the island in 1993. But paradise turned out to be a bit disappointing, and in the end they found life on Tristan too confined, so they left, I think in 1993, with their two daughters. No descendants, so we were back to seven surnames.

Male lineages have also been studied using DNA, this time not using mitochondrial DNA, which is only inherited via females, but looking at the Y-chromosome, which only occurs in males. This study, based on 76 blood samples taken in 1982, carefully avoiding linking results to family names, identified a 'hidden' extra ancestor, of Polish, Russian, or Norwegian origin, probably arriving in the early 19th century (Soodyall *et al.* 2003).

In 2010, a new eighth surname was added to the Tristan families, when Jody Squib, from Dorset, working for the Public Works Department, married Shirley Swain. Whether he will become a new founding father, remains to be seen, depending on what his children are going to do, as in the case of Kelly Green. Here too, time will tell.

There have been many other settlers on Tristan, mostly men, but also a few women, whom I do not cover in this chapter as none of them left descendants. Details can be found in the books of Brander, McKay, Munch, and Crawford. I make an exception for Joseph Fuller. He was a captain from New England who spent many years at sea, sealing, whaling, and 'elephanting' (slaughtering Elephant Seals for their blubber), mostly around Kerguelen (Desolation Island) in the southern Indian Ocean. His reminiscenses were edited and published in 1980 by the American historian Briton Cooper Busch, in his book *Master of Desolation*. In the introduction he tells us that Fuller had a unique background, being born on Tristan da Cunha. None of the books I know about Tristan mentions Fuller as a family name. His father married Mary Ann Glass, the first child born on the island in 1817. It was the first marriage of a true island girl, in 1833, when she was only sixteen years old. The Tristan books say that Mary Ann married (in 1833 indeed) a certain Samuel Johnson. There is no further mention of this man, except that he left Tristan in 1847, presumably because he had seen enough of the island. The conclusion is simple: Samuel Johnson must have been an alias, which so many men adopted in those days, and his real name was Joseph Johnson Fuller. Mary Ann's younger brother Thomas named his son after him: Joseph Fuller Glass (the one who married Agnes Smith).

The Fullers stayed on Tristan 15 years, and six of their eleven children were born there, including Joseph Junior, our future sealer, in 1839. The true story of the family leaving is that Joseph senior developed a lip cancer, and there being no doctors on Tristan, he sought treatment in America, where already quite a few of his inlaws (Glasses) were living in New England, working in the whaling industry. Joseph junior soon went to sea, and worked himself up from green hand to captain. He was shipwrecked on Kerguelen, where he was rescued after almost a year by his uncle Robert Glass, master of the *Francis Alleyn*. After Glass retired Fuller became captain of the same ship, with his brother Moses as first mate, and his cousin Bob Glass (the one who later married Elisabeth Smith) as third mate. They made three voyages with the *Francis Alleyn*, leaving New Bedford in August 1890, September 1891, and June 1893. The 1891-1893 itineraries (perhaps also including the 1890 voyage) have been depicted on the first day cover of a set of new Tristan stamps on the sailing ships, issued in 2019. In October 1890, they visited Tristan. Peter Green wrote about this to his friend, the poet George Newman. Brander quotes this letter in full, the relevant sentence being:

"On the 30th October a schooner came to Tristan. She was an American, from New Bedford, Mass., and the captain and the mate both proved to be natives of Tristan da Cunha. The captain, Joseph Fuller, and the mate, Moses Fuller, were brothers. They left Tristan about 35 years ago, in an American whaler, but they had not forgotten the old 'folks at home'."

Brander adds a footnote to this letter:"It is very likely that Green's memory played him false, and that Joseph and Moses Fuller were J. and M. Taylor, two sons of John Taylor and Elizabeth Glass."

We now know that Green was right and Brander was wrong.

It is amazing how freely we can dig around in the family histories of Tristan. There can hardly be another village in the world where so much information on the people is available to the general public. A possible exception is Pitcairn, the island of the famous mutiny of the Bounty. The people there, like those on Tristan, live permanently under a magnifying glass, closely watched by the outside world. And they too have suffered from a bad press, even worse than Tristan I believe. Well, enough digging for now.

15. A national anthem

The English poet Gerge Newman (1835-1911) was planning to write a biography of his adventurous brother in law, Captain Anderson of the *Sir Ralph Abercombie*, shipwrecked at Tristan in 1870. He wanted to know the details of Anderson's stay on Tristan, where he was hosted by Peter Green, so he wrote a letter to Green. They remained pen pals forever, and Newman became a Tristan addict. In 1900 he wrote a 'national anthem' for the people of Tristan da Cunha. I don't know if it ever reached Tristan. I also don't know whether Newman thought of a tune to which the islanders could actually sing it. Green died in 1902, so he may still have had time to promote his friend's creation, but it did not stick. Today the people on Tristan do not know about it. Most of my British Tristan-fan friends also don't know it, let alone the general public. So here it is:

Wrapt in the solemn sound
Of ocean all around,
Upon our rugged, wind-swept island home;
Far from the busy world,
Th' Atlantic waves are hurled,
Against our rocky shore in mist and foam.

Yet here in peace we dwell,
By ocean guarded well,
And the old flag that flutters in the breeze;
Though circumscribed we be,
We live as near to Thee,
O God, as those who dwell 'mid calmer seas.

Now while our prayer ascends
To Thee, whose ear attends
The single suppliant, or a nation's call;
Oh! do Thou deign to bless

With health and happiness,
Our Queen and keep us loyal, one and all.

And, the wide earth around,
Where Britain's flag is found,
Let Liberty still flourish fresh and green;
And bid us all rejoice,
"And sing with heart and voice,"
The glad refrain, "God save our gracious Queen."

(Other Lyrics, p. 48.)

16. Reverends and teachers

Missionaries have always played an important (sometimes dubious) role in Tristan history. The attitude on Tristan towards them was ambivalent. When they had no missionary, they desparately wanted one, and when they had one they were often greatly relieved when he left. What these missionaries had in common was knowing what was best for the islanders, much better than the islanders themselves. Born meddlers. And people on Tristan don't like that at all. One thing all early reverends agreed upon was that the island was basically uninhabitable, often balancing on the edge of starvation after failing potato crops, and that total evacuation was the only solution.

In the beginning William Glass took responsibility for teaching and lecturing about God. He would marry people, but not baptise their children - that would have been going too far. The first teacher and lecturer from the outside was August Earle in 1824. The second teacher was Benjamin Pankhust in 1830 from a Caribbean island where his father was postmaster. He had a serious falling out with his father and was sent to Tristan to cool off. The schoolchildren were terrified of him, with his wild eyes and unkempt, uncut hair. In 1832 he went home to make peace with his father but when that failed he returned to Tristan, where he stayed a few months longer before disappearing.

In 1834 an advertisement was placed in a South African newspaper asking for a husband and wife team of teachers. There were no applicants. In 1835 Reverend J. Applegate, travelling on the *Wellington*, spent a day on Tristan. He held a church service and baptised 29 children. Similarly, Reverend John Wise, on board the *Augusta Jessie* in 1848, held a service and baptised 41 children. John Wise reported very favourably about Tristan and its people to the Society for the Propagation of the Gospel (SPG), with the result that they sent the first missionary to Tristan for a prolonged period, Reverend William F. Taylor.

Taylor left England on the 23rd of November 1850, with the *Earl of Rippon*. On his way to Cape Town the captain agreed to stop at Tristan. He was prepared to spend one day only and if landing was not possible, Taylor had to continue to Cape Town. On 8th February, Tristan should have come into sight but the captain failed to find the island. In the evening he decided to give up and move on to Cape Town. A disappointed Taylor watched the sunset from the stern, when he suddenly saw a strange little point high in the clouds. That must be Tristan, he said, but nobody believed him. Later with the sun sinking further and the light changing it became obvious that this was Tristan. The captain turned and the next morning Taylor was able to land on the island. He soon started baptising and teaching. He was assisted by Mary Riley, daughter of 'Old Dick' Riley, one of the few people on Tristan who could read and write. She turned out to be an excellent teacher. Church services were held in the Glass house, but the islanders promised to build a real church soon. Taylor was very happy about his flock. But the church was never built. Men were always busy with other things, usually concerning sheep or potatoes. They were always willing to help, but never today, always tomorrow. And tomorrow there would be another tomorrow. Taylor's opinion on the islanders changed from favourable to unfavourable.

During Taylor's stay, the number of whalers visiting the island rapidly declined, as there were no more whales to be found in the area. As a result, many commodities on Tristan became very scarce, and the island was poverty stricken. Taylor was also worried about the surplus of women. Many boys left the island to find adventure at sea. There were at least twelve marriageable young women without any prospect of finding a husband. Taylor saw only one solution: total evacuation. As a godsend, James Glass came to his assistance, son of William. He emigrated to the USA, but came to Tristan to marry his love, teacher

Mary Riley. After the wedding he left again, returning in 1856 to fetch her (including two children she gave birth to during his absence). The entire Glass family took the opportunity to join James and Mary, leaving the island, to settle in America (father William had died three years earlier). With the Glass exodus, 25 people left the island, including William's widow with two unmarried sons and three unwed daughters, and also the deserted Jenny with her two children. In addition, Charles Taylor left with his wife and ten children.

A quarter of the population was gone. Taylor was content, and worked hard to get the rest removed too. He almost succeeded. The bishop of Cape Town, alarmed by Taylor, sent the *Frolic* for the evacuation, shortly after the Glass exodus. Quite a few families were prepared to go. Only the Swains, Cottons and Hagans said they wanted to stay. Peter Green had still to make up his mind yet. The *Frolic* was too small to take everyone who wanted to leave, being eight places short. The whole operation was postponed until a larger ship could be found.

In March 1857 the *Greyhound* took 45 people to the Cape, including the reverend himself. He settled with his flock in Riversdale. He had also removed the birth, marriage, and death registers from the island, and the church bell, which came from the shipwrecked *Blenden Hall*. Taylor thought his job on Tristan was done and was convinced the others would follow soon. Only 28 people stayed on Tristan, the Greens, Cottons, Swains and Hagans. Of course, Taylor's scheme failed. The remaining families happily reproduced and, in addition, sons of Rogers and Glass returned to settle on the island. Within a few decades the population was back to full strength. In 1867 reverend Milner, on the *Galatea*, baptised another sixteen children.

Taking care of the souls could be undertaken by visiting reverends, but schooling was a different matter. After both teacher Mary Riley and Reverend Taylor were no longer there, Peter Green, the most literate man on the island, assumed the task of teaching the children to read and write. In 1870 he got some reinforcement, when Cornelius Cotton brought a wife from Saint Helena, Frances (Fanny), who agreed to live on the island providing her children received a good education. She soon took up that task herself, gathering children in her home on Sundays, where she also read them prayers. Fanny has been described as 'a character of her own', and she disagreed with Peter Green about schooling methods, so they ended up each having their own class of children. Fanny left Tristan after her husband died in the life-boat disaster.

In 1876 the bishop of Saint Helena came up with a perfect candidate, the 24 year old Mr Brady, teacher on Saint Helena. His father was a respected military man, his mother 'almost white'. He was skilled and deeply religeous, but could not get the approval of the SPG, not being a 'real' reverend. Two years later the bishop found another candidate, but he too was disqualified by the SPG. Finally, after a passing captain published a favourable report about Tristan in the *Times*, Reverend Erwin H. Dodgson applied. He was a former missionary in Africa. He had to return to England for health reasons but the dull life in Britain no longer satisfied him. His brother Charles Lutwidge Dodgson was a reverend too, but also a mathematics teacher in Oxford, and, as Lewis Caroll, author of *Alice in Wonderland*.

As usual, it took Dodgson a while to arrive at Tristan. First he got stuck at Saint Helena, where all the whalers had already left. Finally, the SPG chartered a small schooner, the *Edward Vittory*. At Tristan the reverend got ashore dry but his belongings did not. The captain waited too long before unloading and sudden swells threw the ship on the cliffs. The stone baptismal font sank to the bottom, as you may expect a stone font to do, but miraculously it was later found on the beach, undamaged. An act of God.

Dodgson was happy about his flock. He saw no bad

behaviour and heard no bad language. Church attendance was good and the people promised him they would build a real church soon. But, as in Taylor's time, the church never materialised. There were always the sheep and potatoes to look after. Maybe tomorrow.

In 1882 disaster struck, with the arrival of the rats from the *Henry B. Paul* which floundered at Sandy Point. Dodgson urged the men to kill them as soon as they landed, but nobody thought a few of these small animals could cause harm. Within two years they swarmed all over the island. First they feasted on the juicy, fat small petrels, which nested in their millions in burrows on the mountain. But when these were gone the rats entered a period of famine. Desperately they started to dig up the potatoes in the Patches, and climbed the trees at Sandy Point, to get at the last peaches. And in every house they were seen, lapping milk off the table. They caused a deep economic crisis for which the reverend only saw one solution: total evacuation.

In 1884 Dodgson left the island, again suffering from health problems. His evacuation plans caused unrest and discord. Many boys had left the island, joining whalers, and as in Taylors time, there were too many single women. Some families felt like leaving, others did not, and in the end the people stayed. Dodgson no longer thought favourably about his flock. In a pamphlet he wrote that if Darwin was right, the people of Tristan were on their way to evolve into apes. When he left, some of Dodgson's possessions remained on the island. He may have forgotten about his pamphlet, or had mislaid it, but it was accidentally discovered by Peter Green. He was furious and wrote a very angry letter to the Admiralty.

In 1885 bad luck struck again with the life-boat disaster, when almost all the men of Tristan perished at sea. Dodgson heard about it in England and immediately decided to return to Tristan. His flock needed him. In August 1886 he arrived, bringing clothes and food provisions. No rat poison and seed-potatoes that the people had asked for. These were not necessary, he thought, with the evacuation happening soon.

The captain who brought Dodgson back to Tristan reported that the situation on the island was not really that bad. The potato harvest had been good, and the people had even sold potatoes to ships. Dodgson reported the opposite. He also said that the majority had expressed their wish to leave if the Government would grant them a reward of five pounds each. But either this was never taken seriously or the report went astray. In the same period another ship also reported that things on Tristan were alright. Several young men had returned to the island to help the widows and the situation seemed under control. As a result, it was decided that the islanders would be best helped if it was arranged that a ship would visit once a year bringing mail and provisions. The evacuation plan had failed again. Dodgson left in 1889, bitter and disappointed.

After Dodgson left a whole generation grew up without proper schooling. Fanny Cotton had left and Peter Green became too old. Suzanna Swain, widow of Sam Swain who died in the life-boat disaster, taught the children some reading and writing, temporarily asisted by a Mr Cartwright, who stayed on the island for a while after having been shipwrecked with the *Allenshaw*.

In 1903, the British government decided to move on with evacuation without the aid of reverends. The *Thrush* was sent to the island for preparations and to report on the presumably alarming situation on the island. Of course, things were not as bad as expected, but the people promised to seriously consider the question of leaving. So in 1904, the *Odin* arrived at Tristan to take them away. They would be housed in a settlement in South Africa and live happily ever after. But again, the vote went wrong (or right, depending on your point of view), and they did not want to leave. The *Odin* sailed back with hardly any passengers. Seven people left, 71 stayed. The ships's reverend baptised 17 children.

It was evident that the Tristan people who were never going to leave their island urgently needed a new reverend and a teacher. The British government offered to supply a teacher if the people of Tristan would contribute 75 pounds to his annual salary. But they had to decline the offer as their total income for the previous year amounted to just five shillings. However, they were more successful with the reverend. After an advertisement was placed in the *Standard*, Reverend Graham Barrow, who was over fifty years old, applied and got the job. He had a special bond with Tristan. As a four year old child, his mother had survived the shipwreck of the *Blenden Hall*, and had told him many exciting stories about her adventures and the endless hospitality of the Tristan people. She must have been a daughter of the terrible Mrs Lock/Key. So Barrow felt he had to do something in return.

With his wife Kate and a servant, Barrow departed for Saint Helena on the 18th of November, 1905. But like Dodgson he became stranded there. No ships available. Next, in January 1906, they travelled to Cape Town. On April 1st they found a ship, and on April 8th they arrived at Tristan. A strong wind from the northwest was blowing, so the captain decided to land the reverend and his things near Sandy Point, assisted by the islanders and their boats. The sea was rough so unloading was not easy. Many things got wet, including the shivering Reverend, his wife, and servant. The harmonium fell into the sea but, being packed in a zinc-lined chest, it survived, only requiring to be retuned. The place where Barrow and his belongings washed up is now indicated on the Tristan map as Down-where-the-minister-land-his-things.

The Barrows stayed three years on Tristan da Cunha. They were of a different breed than their predecessors and never mentioned the word evacuation. Enthusiastically they took up the task of schooling, Barrow teaching the older children, his wife taking care of the little ones. She wrote a book, *Three years in Tristan da Cunha*, in which she gives very charming and warm descriptions of the children. The Barrows were good sports, making many excursions on the island. The Reverend went to the peak, and even swam in the ice-cold crater lake. He also joined the islanders on a pilgrimage to Inaccessible to the place where his mother was shipwrecked.

But over time, their favourable opinion of the islanders started to erode, as it had with all previous reverends. Again, there was no indication that the promised church would ever be built, and what bothered Mrs Barrow most was the people's indifferent attitude towards their cattle. The Settlement Plain was heavily overgrazed, and reducing the number of cattle was not seen as an option. Nobody could be persuaded to grow hay in summer to provide fodder for the winter. As a result in the winter the poor cattle just starved and died. During their first winter over 300 cattle died, almost half the herd.

While the Barrows were there they were twice visited by a Mr Keytel from South Africa, who had great plans to boast the economy of Tristan da Cunha: agriculture on Inaccessible, exploitation of Guano on Nightingale, wool-export to Cape Town, and exporting dried fish products. Nothing came of it. The fish refused to dry in the humid climate, becoming fly-blown and rotten and the sheep were suffering from scab. After two years hard work Keytel Enterprise went under. Keytel also made a collection of bird skins, which is still in the South African Museum of Natural History in Cape Town. No Moorhens because they were no longer to be found.

The Barrows left Tristan in 1909, intending to come back after a short period of leave. For months they waited in Cape Town for a ship to the island, and after that they tried for some months to find transport from Rio de Janeiro, but in the end they had to give up. They returned to England.

After the Barrows left things around Tristan became very quiet. Sailing shipping died out, and all sea traffic from Europe to the East went through the Suez Canal. In 1920, for instance, no ship came in sight at all for the whole year. But the year before, a huge steamer, with four chimneys, the *Yarmouth*, visited Tristan. Reverend Turner spent half a day ashore, and baptised another 15 children. The visit of the *Yarmouth* was the result of the efforts of a private benefactor, Douglas Gane from London. Gane visited Tristan in 1884 on his way to Australia and was shocked by the poverty he saw. A year later he heard about the life-boat disaster and decided to do something about it. He started raising funds and established the Tristan da Cunha Fund in 1916. In 1926 the fund was made official, when the Royal Geographical Society and the Royal Colonial Institute were appointed as trustees. In 1961 the activities of the Fund were suspended when the Tristanians were evacuated after the volcanic eruption. Even though many people returned, the Fund ceased to exist in 1964. The people of Tristan da Cunha were supposed to be able to keep their own economy afloat.

Back to the difficult 1920s. After a cry for help, Reverend Rogers was sent to Tristan by the SPG. Henry Rogers was 42 years old, his wife Rose only 19. Like Kate Barrow, Rose Rogers kept a diary, which she publised as *The Lonely Island*. Rogers saw in the newspapers that famous explorer Shackleton, with the *Quest*, would visit Tristan da Cunha on the way south to Antarctica. Rogers asked for a passage but his request was denied. Not enough room on board. But Shackleton was willing to deliver his letter, telling the people they were on their way. With the usual delays, the Rogers' arrived at Tristan in April 1922, with the *Tacoma Maru*, sailing from Cape Town to Buenos Aires, the first Japanese ship ever at Tristan. Whilst on board, the reverend and his wife learned how to eat with chopsticks.

For the islanders the arrival of Rogers was a total surprise. Two months later the *Quest* delivered his letter announcing that they were coming. When the boats came out to meet the ship, Rose was struck by the appearance of the men. They looked a sorry lot, unshaven, wild-eyed, in patched-up clothes, wearing moccasins or no shoes at all.

The men were shouting and gesticulating wildly and she was frightened because she was afraid of dark races. But later, when her husband had explained why he was there, they all took off their caps, raised a hand, and gave three cheers. Her fear dissipated, and instead of being afraid of them, she felt pity for these poor fellows.

The reverend and his wife found Tristan in a deplorable state, in perhaps the worst of all bad times the island had so far experienced. There had been no ship for more than a year, and there was a shortage of everything. During their three years' stay things only grew worse. Out of three potato harvests, two failed, and many families were on the verge of starvation. They gave away as much as possible but ran out of their own provisions too. Everybody lost weight, including Mr and Mrs Rogers.

Hunger was temporarily relieved during the egg-laying season of the seabirds. People would go to the other side of the island, or to Nightingale and Inaccessible, to collect thousands of eggs. For weeks on end, they would eat nothing but eggs. Rose writes:

"Penguin eggs were said to be used in immense quantities in September and October. 25,200 eggs were said to be used in one year, and 7,400 eggs have been collected in one day by boats at Stony Beach, Trypot, Seal Bay, and Sandy Point rookeries. Black eaglets are got in June and July, and we found these good eating, but molly mawks, which are hunted from January to March, were very strong and unpleasant. 2,139 mollies were taken in one January and 4,800 in March. From these figures it is evident that people would be starved if they had not sea-birds and their eggs to fall back upon as articles of diet. The people eat every conceivable kind of sea-bird except penguins and sea-hens, but petrels, molly mawks, and black eaglets are most liked. Night birds are getting scarce, and albatross have left the island entirely. All varieties of eggs are eaten, but most are fishy and indigestible. I tasted six varieties of sea-birds' eggs, but black eaglets, penguins, and night birds are the best, and molly mawks the worst; petrel eggs are not bad eating. They all are best fried hard on both sides, and eaten with plenty of salt and pepper."

Like Mrs Barrow, Rose found the islanders very friendly and hospitable, but it was extremely difficult to get anything done. They were always willing to help, but never today, always tomorrow. In her eyes, they were just like a bunch of big children. She also found their hygiene wanting. They would butcher a sheep close to their houses, leaving the entrails laying around for the dogs. Their watercourses were polluted, and many were suffering from roundworms. Every house was infested with fleas. Rose acted as a doctor, bandaging wounds or broken limbs. But people sometimes feigned injuries to use the bandages for patching up clothes, or as cooking cloth to boil potato puddings. She also discovered that her clove-based tooth-ache medicament was used by the girls as a scent for dancing parties.

Surprisingly, Rogers managed to have the church built, at last, after his predecessors had tried in vain for three quarters of a century. He also installed an island council, which could make decisions for the benefit of the island, and improve living conditions. Furthermore, he founded a boy-scouts group, the Penguins. With his support, the islanders sent a request to the Admiralty, asking if a ship could visit them once a year. The response was negative. Poverty on Tristan would continue for more years. Of course, inevitably, evacuation plans surfaced again. To look into that, the *Dublin* was sent to Tristan in 1923. The *Dublin* was loaded with presents, thanks to the Douglas Gane and the Tristan da Cunha Fund, so the evacuation was no longer seen as a necessity.

Rose had her first child, during her stay on Tristan. She asked the islanders to give him a name. They chose Edward, after the Prince of Wales. In 1925, Henry, Rose and Edward left Tristan. In 1926 Rogers received a cry for help. The economic situation on Tristan had further deteriorated and without a padre life was unbearable. Henry decided to return to the island, but two weeks after he made that decision he died of general weakness, most likely the result of the hardship during the preceding years on the island. In a lecture, given shortly before his death, he said:

"The rushing of the Atlantic waves is still in my ears, while I can still smell the kelp on the beach, and can see those eager, tear-strained faces pleading for our return - for my heart is ever there and I hope I may yet be able to return to the little island flock I love so well."

After 1926 conditions improved. The Tristan da Cunha Foundation obtained official status, and there was a growing number of cruise ships visiting, leading to an increasing influx of goods and presents. When the economic crisis struck Europe in the 1930s, the Tristanians had just survived theirs, and saw better times.

Between 1927 and 1929, there was reverend Pooley. I know no interesting stories about him, which probably means he was a quiet and nice man. He brought an assistant, the theology student Philip Lindsay, who mostly occupied himself with the young people. After Pooley, Reverend Partridge came to the island, a different character. He revealed himself as a true potentate. He appointed himself as magistrate with absolute power, introduced a judicial system and used the cane to punish sinners. In his pursuit of moral values he found an ally in Mrs Francess Repetto, widow of the shipwrecked Andrea Repetto, and granddaughter of Peter Green. When Peter Green died in 1902, Andrea was given the role of headman, but as such, he remained rather inconspicuous. When he died in 1911 his son Willie took over, becoming known as 'Chief Willie'. But in reality it was his mother Frances who was the leader as head woman, a status Partridge made official. Frances is generally described as highly bright, intelligent, and witty, with a good sense of humour, literate, and with high moral standards, one of the most outstanding personalities Tristan had ever known. Yet, I am not so sure, if I had known her personally, if I would have liked her that much (or she me). She had the habit of visiting houses by surprise, to check whether the housewives did their job properly and kept their dwellings clean and tidy. The worst of all, and the laziest mother was called 'Long Lena'. She hated the inspections by Frances Repetto, and once even showed her a fist. Partridge's court found her guilty of laziness and negligence and she was denied access to the church for a while. Partridge fabricated a pillory from old cartons, a stock, in which Long Lena was exposed to the public for a whole day.

Partridge was succeeded by Harold Wilde, possibly an even worse potentate. He found Partridge's pillory an excellent idea, and erected a permanent one himself, which has never been used. Like his predecessor, he wanted to develop the useless islands of Nightingale and Inaccessible and start agriculture in a rat free environment. Wilde was a robust, extrovert man with a loud voice, who liked to slap people on the back or on the shoulder, a habit the islanders did not like that much. On the one hand, Partridge and Wilde were deeply respected, on the other hand, they evoked resistance, leading to a growing number of people turning their backs on the Anglican Church, to become catholics.

In the Second World War Tristan was occupied by British soldiers for the second time, to keep an eye on German activities in the Southern Ocean. In addition, a weather station was established. This meant the constant presence of a team of expats, who were housed in what became known as 'The Station'. The naval chaplain, Mr Lawrence, like Partridge and Wilde, had great plans for the development of the Tristan economy. He saw a large potential in the exploitation of the tremendously rich resource of the crayfish (or crawfish as the islanders called them), now known as the Tristan Rock Lobster. He envisaged two canning factories, one in the Settlement, and one at Sandy Point, connected by a permanent road or even a railway. He got the Lamberts Bay Canning Company interested, and they organised an exploratory expedition with the *Pequena* in 1948, with Lawrence as expedition leader. It also included three marine biologists, a fish oil expert, and a medical pathologist. The expedition came back with very favourable reports, and that is how the lobster fishery started, which is still the main source of income for the island today.

In the beginning the relationship between the company, the administration, and the islanders was complicated. The people were not very good at keeping to working hours. Regardless of contracts, they would leave their work any time they liked, to look after their sheep, or to dig up potatoes, often only showing up for just a few hours a day. The islanders used to go to Nightingale three times a year. First to collect guano, then to harvest eggs, and finally the 'fatting trip', extracting oil from thousands of fat petrel chicks. During these trips the workforce on the island would be diminished. Eventually, Company and islanders arrived at a compromise, where the islanders would have the exlusive right to fish around the main island in their small boats on so-called 'fishing days' (with calm enough seas, happening about a hundred times per year), while the company ships would fish around the outher islands, including Gough. Between fishing days, the people could look after their sheep and potatoes, and the Nightingale trips continued. This is still the situation today. The second factory at Sandy Point was never constructed. Neither were the road or railroad.

With the permanent presence of so many expats, it was necessary to separate religion and schooling. Also, the British Government wished to formalise the political status of Tristan by appointing an Administrator. The first was Hugh Elliott, who came to the island in 1950. Medical staff and a teacher also arrived. Towards the end of the 20th century this situation has changed. Administrator, doctor and reverend are still outsiders, but nursing and teaching have been taken over by islanders.

About the post-war reverends I have no interesting stories to tell. I only want to mention Patrick Helyer, who served on Tristan in the late seventies. It was his last post, and after he retired he went to England, where he lived in a house called 'Tristan', where I visited him in 1993. His house was adorned with many pictures from Tristan, and he had a nice collection of old books about the island, some of which he sold to me. He used to send piles of Christmas cards to Tristan, which he mailed in August, to make sure they would get there in time. Helyer also assembled more than 3000 titles of publications which mentioned Tristan. Books, book chapters, articles in magazines, scientific papers, newspaper cuttings, and the like. With the help of Michael Swales, his bibliography was finally published in 1998. Patrick Helyer died in 2013 at the age of 97.

After the notorious Father Wilde left, many catholic 'converts' returned to the Anglican Church. The bishop of Cape Town promised the catholics that he would send a permanent priest if there was fourty of them. It never happened. What they did get, though, was their own church, which was built in 1998, with a real tower and a beautiful stained glass window, designed by an Australian artist. A casual visitor may think this is 'The' church of Tristan, as the modest Anglican Church of St. Mary, serving the majority, is much less conspicuous. It is the same everywhere in the world, with catholic churches being much larger and more pompous than the protestant ones, a fact my anti-papal mother used to complain about.

17. Atlantisia

Sir Ernest Shackleton's voyage with the *Quest* was a disaster. He wanted to go to the least known parts of the Arctic, in the Canadian Archipelago, and he promised his wife that he would bring her a Polar Bear skin. He had already obtained some funding and was waiting for a large contribution from the Canadian government. When that did not materialise, he decided to turn south again, instead of heading north. He had obtained a Norwegian wooden-hulled sealer, *Foca*, which had to be refitted, and was renamed *Quest* by his wife, Lady Emily Shackleton. Hubert Wilkins was on board as photographer and naturalist. He knew about the extinct Island Cock of Tristan da Cunha, and in the Challenger reports he had read that a similar, as yet undescribed, bird was living on Inaccessible. The plan was to visit Tristan on the way south (and, among other things, deliver Rogers' letter to announce his arrival before he got there himself). Engine trouble began when they were only in the Gulf of Biscay and the ship had to go to Lisbon for repairs. Near the Brazilian island of Fernando Norhona the engines failed again, and instead of heading for Tristan they had to divert to Rio de Janeiro for a thorough check-up. Now it was too late in the season to go to Tristan before going to the Antarctic, so it was decided to leave Tristan for the return voyage, a year later.

The tensions about funding had caused Shackleton to suffer a mild heart attack in London. Whilst in Rio he had a second one. On January 4th 1922, the Quest arrived in Grytviken, the Norwegian whaling village in South Georgia. That night Shackleton had his last, fatal, heart attack. He died the next morning. His body was taken to South America, to be shipped to England, but his wife did not want him there and decided that he should go back to South Georgia. He was buried in the cemetary of Grytviken on March 5th 1922. His friends did not attend the funeral as the *Quest* had already left to continue her Antarctic expedition.

At the end of the Antarctic summer, the *Quest* finally arrived at Tristan. The next day they sailed to Inaccessible, where a landing was not possible. Instead they went to Nightingale, where Wilkins found the large-billed bunting, which the islanders called Big Canary, but which was not yet described by a scientist. It was named after him: Wilkins' Bunting *Nesospiza wilkinsi*. Later, on the same day, they managed to land on Inaccessible, but Wilkins failed to find the little Island Cock. Perhaps he tried too hard. Instead of looking for them he should have waited for them to look for him. When you sit still they like to come out of the vegetation, to give you an inquisitive look. When I (at last) saw mine, the bird was just as curious to look at me as I was to watch him.

Wilkins left instructions how to prepare bird skins with Henry and Rose Rogers. In February 1923 Rogers managed to obtain two specimens, which were sent to London. Dr. Percy Lowe described them as a new species. It was one of the strangest birds he had ever seen, with it loose, fluffy feathers looking like a two-days-old chick, just as the Stoltenhoff brothers had told Wyville Thompson. He named it *Atlantisia rogersi*, in honour of Reverend Rogers, who had secured the specimens.

The German scientist Alfred Wegener had just recently launched his idea of continental drift, matching the coastlines of the Americas and the Old World. Not many people believed him yet, certainly not Percy Lowe. He also held the firm belief that in the course of evolution, flightless birds were more primitive than flying ones, and the idea that flighlessness could be a secondary condition, was totally unthinkable. Combining the two disbeliefs, the little rail must have reached Inaccessible on foot, walking all the way from South America. The bird was the ultimate proof that the now sunken continent Atlantis, connecting South America with Inaccessible, had really existed. Hence the name *Atlantisia*.

117

18. Wooden shoes

Dutch hisorian Jan Brander was born on July 24th 1879, into a family of farmers and bulb cultivators, north of Amsterdam. In 1910 he moved to Vlissingen, where he taught geography at a secondary school. His main interest was the history of Dutch whaling. In 1933 he published a book on the island of Jan Mayen, in the arctic North Atlantic. After the Dutch had killed most of the whales around the island, they totally exterminated the Walrusses for their valuable ivory tusks. They would simply chop off the frontal part of the upper jaw, including the tusks, without first killing the animals. Around Spitsbergen, Walrus populations are slowly recovering from this massacre, but on Jan Mayen they never returned.

After this island in the north, his interest turned to an island in the south, Tristan da Cunha, after he read an article about Pieter Groen in a geographical journal. He dug into Groen's past, and published his findings. He started a correspondence with Groen's (Green's in the meantime) granddaughter, Frances Green, studied all the archives he could find, and eventually, in 1940, published his book *Tristan da Cunha, 1506-1902*, by far the best and most detailed book on the early history of Tristan. All later authors, including myself, happily copied him.

With great interest, Brander followed the voyage of the submarine K-18 in 1934. In those days the Dutch were very proficient submarine builders, and were very eager to sell their product to other countries. The K-18, brand new, equipped with the latest findings, was going to be repositioned to the naval port of Soerabaya in the Dutch East Indies, and on the way, she would pay a visit to Tristan da Cunha.

The K-18 had four special assignments during the voyage. First of all, it had to be a promotional trip. So there were embassy parties, and dinners with captains of industry, in Dakar, Recife, Rio de Janeiro, Montevideo,

Buenos Aires, Cape Town, Durban, Mauritius, and finally Fremantle in Australia. The second assignment was science. The famous professor Vening Meinesz was on board. He was a specialist in measuring gravity. Alfred Wegener's idea of continental drift was only a decade old, and still very controversial. Gravity measurements had been taken all over the world, except over ocean floors, and that might give a clue, the professor thought. The problem was that you measure gravity with a pendulum, as the frequency of swinging solely depends on the length of the pendulum and gravity, but a pendulum does not work on a rolling ship. For the same reason, time keeping has been a major problem at sea for centuries, making establishing the longitude at sea very difficult, resulting in the loss of many ships and lives.

Vening Meinesz solved the pendulum problem by diving deep enough to get away from the influence of wave action. He was the first to map gravity over the Atlantic sea floor and in the Indian Ocean. The third task of the K-18 was to act as a radio beacon in the mid-Atlantic, coming up in the right place and at the right time, for KLM's first trans-Atlantic flight, from Amsterdam to Paramaribo, capital of the Dutch colony of Surinam in South America. The contact was made, but it was a cloudy night so the crew were disappointed not to see the aeroplane. The fourth assignment was visiting Tristan da Cunha, a special request of the British government, because there had been (again) alarming reports about the health and welfare of the poor people on their wretched island. In all likelihood, total evacuation was considered to be the solution.

On March 21st the K-18 arrived at Tristan. The people of Tristan had never seen a submarine, and some children thought they saw a very strange whale. Most men were on a trip to collect eggs on Inaccessible, so it took a while to find enough hands to fill a boat to meet the K-18, Reverend Wilde taking the lead. Along with the mail, some crew members came ashore, including Lieutenant

Wytema, who later published a book about the voyage. He noticed the poverty on the island, but in spite of that, he found the people cheerful and in general very healthy. And he noted, like many others before him, that they all had exceptionally fine teeth, even the oldest people. No sweets, no candy. The only health problem he noted was a nasty cough amongst many of them. This had been known for a while, and was generally attributed to the lack of proper footwear. Their moccasins were not waterproof and in the rainy climate they constantly had wet feet. Reverend Wilde knew the solution. He slapped Lieutenant Wytema on the shoulder and loudly exclaimed: "What these people need is decent Dutch wooden shoes to keep their feet dry!" But he doubted if the Tristan da Cunha Fund would be willing to buy clogs in Holland because of of the Buy British policy. The reverend suggested that perhaps another solution could be found.

Jan Brander got the message. He found the other solution. He started to raise funds through local newspapers, and soon he had assembled a pile of 750 wooden shoes and there were more to follow. The first batch, 273 pairs of clogs, arrived on Tristan in 1937, with the *Cap Pilar*. In 1938, HMS *Milford* delivered the rest, an enormous quantity in large sacks, more large sacks, and even more large sacks. It was great fun to go through the pile and find a pair that fitted. Choice was limitless, as there were far more pairs of clogs than people.

Jan Brander had more ideas. In the 1950s, he thought that it would be nice to have a Dutch cheese for Christmas, on every table in Tristan. He raised funds again, and bought a large quantity of Edam cheeses (these are round, a bit smaller than a football, coated in bright red parraffin wax, which distinguishes them from the more widely known yellow wagon wheels from Gouda). The cheeses were shipped from Rotterdam to England, awaiting further transport. But alas, in a London warehouse, they went mouldy and rotted away. They never reached Tristan.

Jan Brander died in 1964 in Vlissingen. In spite of his

adventurous mind, and his great interest in remote islands, he never ventured to travel himself. As a teacher, he was known to be strict, not tolerating foul language. He kept extensive files on Jan Mayen and Tristan. The Tristan files were donated to the Royal Dutch Geographical Society (KNAG). The Jan Mayen files came into the hands of the retired Captain Albert Veldkamp, a former pupil of Brander. Later, the Tristan files were moved to the National Archives in The Hague. Both institutes keep a list of transferred files, but the names of Tristan or Brander do not appear in either of them. Brander's Tristan archives are just as lost as the moorhens of Dugald Carmichael. Of the wooden shoes, one pair remained in The Netherlands. Jan Brander gave them to his daughter Marietje.

19. Allan Crawford

Allan Crawford never intended to go to Tristan. With a technical education, he worked in several ports in Britain. Ports look out at far horizons, so in 1937 Allan sailed to South Africa to try his luck there. On board he met Erling Christophersen, who turned out to be the leader of a Norwegian scientific expedition to Tristan da Cunha. Apart from conducting scientific research, Christophersen wanted to produce a good map of the island, but a Norwegian making a map of British territory was rather a sensitive matter. Christophersen suggested it would be nice if Allan could do the job, to which he happily agreed. He followed a crash course in geodesy and cartography in Cape Town and armed with a heavy tripod and theodolite, he joined the Norwegians on the German freighter *Anatolia*, bound for Tristan on November 30th, 1937.

Nobody from the outside world has seen so much of the island as Allan Crawford. He camped all over the place for days on end, usually with one or more island guides, mapping every cliff, every gully, every hill. Nobody has been to the peak as often as he did. Crawford became totally addicted to the island and its people.

The Norwegian expedition was the first to carry out research on the island for a prolonged time. So far, there had been the brief visits of botanist Du Petit Thouars in 1793, and Dugald Carmichael collecting Moorhens and other birds in 1817. Also, the more serious expeditions of the *Challenger* in 1873 and the *Quest* in 1923 spent no more than a few days. The Norwegians stayed four months.

In 1933, the Norwegian whaler Lars Christensen brought home a collection of plants from Gough Island, which ended up at the Botanical Institute of the University of Oslo. The botanists found so many interesting plants, that they wanted to organise an expedition. But instead of Gough they chose Tristan, where they could study not only the plants, animals and rocks, but also the legendary people in their peaceful community without crime, without police, without punishments, without diseases, and often without any resources to support their lives. They were seen as fodder for psychologists, sociologists and medical people. Lars Christensen gave financial and logistic support, but did not join himself. After four years of preparations, the expedition sailed to Cape Town, meeting Allan Crawford on the way. There were eleven Norwegians, one British, and a South African. Yngvar Hagen was the bird man, Peter Munch the sociologist, and Sverre Dick Hendriksen and Per Oeding the medics.

The *Anatolia* approached Tristan on the 6th of December in thick fog. Suddenly the mist lifted, and a huge, impressive black wall loomed up, which was a total surprise for the expedition members. But then, around the corner, they saw the friendly looking village on a fresh green slope. Soon the island boats approached the ship. The islanders did not know the expedition was coming, because, strangely enough, nobody had taken the trouble to write to Tristan, and nobody asked if they would be welcome at all. But the islanders were very friendly, and soon helped to unload the enormous amount of equipment, never asking for payment of any sort.

Munch studied the islanders in their daily life, with all their subtle interactions, and on all his visits he had to drink many cups of tea. Often he would be given fresh milk. Drinking it was a problem, not because he did not like milk, but because of the hundreds of flies crawling all over his cup and often falling into the milk. He also noted that there were fleas everywhere. The islanders were very fond of music and dancing, organising parties for every possible occasion, such as Christmas, Easter, and 21st birthdays. Girls going through the village with a plate covered with a clean cloth to bring goodies to another family, was a common sight. Munch gave a description of the traditional pillow dance, where a man grabs a pillow and dances with it to the music. He approaches a lady, deposits the pillow at her feet, both kneel on it and kiss each other on the lips. The lady picks up the pillow, and dances away with it, the man following. Then she chooses a man, repeating the procedure, and so on, alternating between adding a man and a woman to the growing snake, until everybody is involved. The Norwegians had to take part in this.

Hendriksen and Oeding found, like many before them, that the islanders were all very strong and healthy and that they all had impeccable teeth. There were no infectious diseases, apart from a quick cold passing through the population every time a ship visited. The people were lean, almost skinny, but well built. Women never died in childbirth, and stillborns were very rare. Midwives did not do much more than let nature take its course. The only serious medical trouble they had was a nasty cough. Hendriksen and Oeding concluded this was a hereditary problem, and through interviews and earlier reports, they managed to trace it back to two women from Saint Helena; Sarah Jacobs, one of the first five, and Susannah Philips, who came later. They found no ill effects of inbreeding.

The geologists amused themselves with the various kinds of volcanic rocks and ash layers, the marine biologists found many interesting creatures in the sea, and Yngvar Hagen studied the birds. On the Base he was startled by eerie cries from the dark and misty ravines, until he found out that the sound came from the Peeo, or Sooty Albatross. Hagen spent a lot of time on Inaccessible and Nightingale, and he was the first to make estimates of population sizes of the penguins, petrels, and albatrosses. He also gave his attention to the land birds. Island Cocks were no longer there, and none of the islanders, even the oldest ones, could remember ever having seen one. Hagen was also the first non-islander to see the diminutive Flightless Rails of Inaccessible alive. He found it very difficult to see them. He could hear them all the time, squeaking under dense vegetation, sometimes only centimetres away from his fingers, when he tried to pry open their tunnels under the grass. Hagen was also the first to ever obtain a photograph of the little creatures.

The Norwegians stayed four months on Tristan. On March 29th 1938, they left on the Norwegian ship *Thorshammer*. Allen Crawford had left a few days earlier with the *Milford* (which brought the clogs), preferring to go to South Africa instead of Norway.

In 1942, with the military occupation of Tristan, a weather station was established. It was all very secret, and the operation was called 'Job 9'. The job was assigned to the South African Air Force, where nobody knew much about Tristan. They found Allan Crawford, who was added to the crew in the lowest rank, as 'air mechanic'. It soon appeared that the leading sergeant was unfit for the job, so he was fired and Crawford was given his place. A quick promotion.

After two years of secret preparations, the expedion left Cape Town on March 20th, 1942, arriving on Easter Sunday. Island boats approached them, dancing on the waves. Crawford recognised 'Chief' Willie Repetto, and called out his name. "Who zat?" the others asked. "That's Crawford," Willie said. It was a warm reunion. Crawford had the intention of doing more that establishing the weather station and the naval station. He realised that from this moment on, the island would become less isolated, and that the people could not escape modernisation. They still walked around in clothes that appeared to come from a century earlier, and for transport they only had simple oxwagons and their self-made fragile boats. Soon these things would change, so if he wanted to register all that on film, he had to do it now. Especially for this occasion, he had bought a professional 16 mm film camera, and a large quantity of colour films (Crawford's films have since been transferred to video).

With the arrival of the meteorologists and the Navy, time in Tristan became confused. The Navy was kept on Greenwich time, via radio, but the meteorolgists used zone time, based on the fifteenth meridian. Then there was 'Island Time', according to the islanders the only correct one. In the past, a British warship had fired a cannon at the moment the sun reached its highest point, establishing 'Tristan Noon'. Various islanders built a simple sundial in their front yard. Finally there was the 'Crazy Time' of Reverend Wilde, who organised unpredicted church services any time he liked.

Naval Commander Wooley, by order of the Colonial Office, appointed an Island Council, consisting of ten respected islanders, the doctor, and the reverend. Crawford was added as 13th member. The council organised periodic 'ratting days'. With dogs, all stone walls in the villange and around the Potato Patches were inspected, the rats chased out and killed. The tails were collected, and a prize was given to the man with the most tails. Ratting days still exist, and there is now not only a prize for having tne most, but also for the longest tail.

A local army unit was also formed, the Tristan Defence Violunteers (TDV). There were thirteen old-fashioned .303 rifles, four guns from before World War I, a few boxes with hand grenades, and some ammunition. The marines would take care of the guns. Sixteen people volunteered for the TDV, but since there were only thirteen rifles three of them had to be content with other jobs.

Crawford revealed himself to be a creative person. He designed decorations for the TDV uniforms, which he later had manufactured in Cape Town, and he drew up a certificate for them, adorned with island symbols such as a spouting whale. The doctor signed the certificates. Copies were sent to the Chief Commander in Simonstown. The certificate read that in case of actions against the enemy, the holder should obey orders from the signer. But medical officers are not allowed to do that in war time, so the certificates had to be destroyed. Crawford, the doctor, and the TDV had violated the Geneva Convention. Cape officers drew new ones, erroneously taking Crawford's spouting whale for a palm tree. Crawford also brought a stencil machine from Cape Town, and started to produce the *Tristan Times*, which still exists.

After the German capitulation in 1945, the military went home, but the weather station, which had proved to be very useful for the predictions in South Africa, remained. In 1946 it was tranferred to the South African meteorological service. Crawford was appointed chef, so he went back to Tristan for the third time, staying until 1948.

Meanwhile, Tristan had its own postage stamps, another creative activity of Crawford. Earlier, mail from Tristan was stamped, but without postage stamps. At the receiving end a stamp would be added to be paid by the addressee. Envelopes from that period are now valuable collectors' items. Crawford realised that having their own stamps and selling first day covers to philatelists, would generate a welcome income. In 1946 he designed a set of ten stamps, with values from half a penny to two shillings. In the end, only the 1 penny stamp (1d) was printed. A problem was that the islanders had no money to buy stamps, so Crawford added a value expressed in a number of potatoes, which the islanders could use to buy them. On the side of the 1d stamp is the note 'local value 4 potatoes'. After the Navy left, the reverend was the official representative of the government, so Crawford let him sign an official petition asking to issue the stamps. They have been used, even though they were only officially recognised as late as 1979. By then, they were all sold out, of course, being followed up by dozens of new series. Potato stamps are now rare and much sought after by collectors.

In 1948 the South African government asked Crawford to organise an expedition to Prince Edward and Marion, two windy and rainy islands in the southern Indian Ocean, which had not yet been claimed by anyone. The operation took place in greatest secrecy, under code name 'Operation Snoektown'. Crawford went there from Tristan, taking six islanders with him. The South African flag was planted on Prince Edward, and Crawford established a weather station on Marion. The island was infested with mice, so in 1949 five cats were released to control them. But the cats, of course, took to killing seabirds instead of mice. By 1977 there were an estimated 3,400 cats on the island, killing half a million petrels per year. The cats were successfully eradicated between 1982 and 1991. The mice are still there.

In 1955, Crawford came to Tristan for the fourth time, as Honorary Welfare Officer for the island, joining a British expedition that would go to Gough Island (next chapter), and, still as Welfare Officer, he paid a brief visit in 1958. In 1962 he was back for the sixth time, with an expedition of the Royal Society and a few island guides. Tristan was uninhabited at that time, because the people had been evacuated after the volcanic eruption in 1961 (see chapter 22). It was thought that Tristan would never be inhabited again and so, with great sadness, Crawford signed the island, painting his initials on a big rock with white paint: 'ABC 1937-1962'. The inscription is still there, but only the elders know what they mean.

Tristan did not stay uninhabited. The people returned, and Crawford played an important role in the process. He really had become a 'godfather' for the island, like Douglas Gane before him. In 1987 the Tristan da Cunha Association was founded, with Allan Crawford as president. Members are old Tristan veterans, such as ex-administrators or their widows, schoolteachers, reverends and a few islanders, as well as a collection of Tristan fanatics (like myself) and stamp collectors. At the annual general meeting in 1995 Allan Crawford resigned as president, to become honorary president for the rest of his life. To mark the occasion, he showed the colour films he took in the 1940s.

Crawford always stayed creative. In 1980 he designed another series of Tristan stamps, and he has written several books about the island. His map, made in 1938 is still the official, accurate map of the island. In 1996 a seamount (volcano rising from the ocean floor, not reaching the surface), about 400 km southeast of Tristan, has been named after him. Crawford died in 2007.

20. Gough Island

In 1955 a group of young academics from Cambridge, just graduated, wanted to organise an expedition. They initially planned to go to Edgeøya, a bleak arctic island in the eastern part of the Spitsbergen archipelago. But when they visited the Scott Polar Institute, looking for funding, it was suggested that they head south instead of north, to Gough Island, one of the least known islands in the world, with a very rich fauna and flora, but without a decent map as yet. The idea appealed to them, so that is what decided to do.

The leader was John Heaney, an engineer who was 24 years old. He grew up in the mountains of Kashmir and had joined an expedition to South Georgia a few years earlier. Roger Lemaitre, geologist and mountaineer, at 21 years old, was the youngest participant. The eldest was botanist Nigel Wace, aged 26. The other members of the expedition were Michael Swales, zoologist for the larger animals (birds and sea mammals), Martin Holdgate, zoologist for the smaller beasties, Philip Mullock as radio telegraphist, and Robert Chambers as 'general scientist', with a special knowledge of expedition equipment. Robert and John had been to school together and were climbing friends. They planned to go to Gough for six months, without salaries, otherwise the operation would have been too expensive. They found a small amount of funding here and there, and their budget was finally supplemented with a personal contribution from Prince Philip.

Before departure John fell ill, and left the leadership to Robert. In John's place James Hall joined. He had a medical background. In Cape Town another expedition member was added, the South African meteorologist Van der Merwe, to investigate if looking at depressions at Gough would be a good predictor for the weather in the Cape Province. The expedition seemed cursed. One day after arriving on Gough, Robert had to be evacuated because of a back injury. Martin became the new leader.

The expedition first travelled to Cape Town. After a week they joined the naval ship HMS *Magpie*, which was taking cargo and various passengers to Tristan. They would need to spend about ten days in Tristan, awaiting transport to Gough with the fishing vessel *Tristania*. Ten days holiday on Tristan was not a punishment. They made several excursions, climbed the peak, and tested all their equipment. Michael counted Mollies, penguins and seals, Nigel collected plants and took a peat core for later pollen analysis. The ten days turned into six weeks. At last, on November 14th 1955, they set foot on Gough, where they first built their hut in the only place with a safe landing, and where a valley led up into the interior, the Glenn.

When settled, the boys started their work, an important task being to make the first detailed map of the island, dragging a heavy tripod and theodolite all over the island, often in bad weather, and always through difficult terrain. They had a small boat, which they used on calm days to explore the coastline, which looked just as inaccessible as Inaccessible. Vertical cliffs, rising to almost 900 m, with wheeling Yellow-nosed and Sooty Albatrosses, and only occasionally a small beach where landing was possible, between the crowds of penguins and seals.

Climbing up through the Glenn, they reached the upper parts of the island, with rugged peaks and an undulating plain, dotted with nesting Tristan Albatrosses. From a distance the plain looked easy walking terrain but that proved to be an illusion. There were obstacles everywhere, and the going was very difficult because of the many petrel burrows in the soft, peaty soil, which were invisible under the vegetation.

Michael found that the Moorhens were far less scarce than he first thought. They could be heard calling all around in the higher, more dense vegetation. That was the case in all the lower parts of the island, in the Glenn, and around their house. But they were elusive and seldom showed themselves. The best way to see them was just to be doing something entirely different. At first, most observations were made by the non-birders. Just sitting still to let the inquisitive bird come out to have a look at you was the best method. Michel failed to find their nests, and it caused a great deal of frustration when a parent with two small fluffy black chicks walked past their house, just below the window. It must have had its nest within a radius of perhaps ten metres. Later, the same thing happened again.

Michael collected a lot of animals, which he kept in his zoo, to take back to England, alive for the zoo, or dead for the museum. His zoo contained quite a few Moorhens. After all, they could not fly, and with some stealth could easily be caught. That was best done with eight people, surrounding the bird, and then slowly closing the circle. And when the desperate animal finally tried to break through the cordon it could be grabbed.

In May 1956, the expedition was picked up by the South African ship *Transvaal*, which took them to Cape Town, via Tristan, zoo included, Michael's Moorhens did well in the London Zoo, and bred happily in captivity. Some managed to escape, and started to interbreed with the local Common Moorhens. I don't know if they produced fertile offspring, and whether there is still Gough-blood in the British population.

The weather station of Tristan was moved to Gough Island, making use of the expedition hut in the Glenn, but later a new station was built further east, on a higher plateau. The Glenn was a perfect landing place, but with steep slopes on either side, it was unsafe to land there with a helicopter. The new location was on a flat plateau, where a safe helipad was established, but there was no landing place. Heavy materials which could not be flown in, had to be hoisted up to the top of the cliff with a crane from the boats below.

In the publications about the expedition, there is no mention of eight Moorhens that Michael Swales left behind on Tristan, releasing them east of the Settlement.

21. Crisis in Utopia

In 1961, all the reverends and the government, with a little help from God, got what they had wanted so badly for more than a century: total evacuation of Tristan. The government had no plans to let the people ever go back to their island.

It started on Sunday August 9th with a short but heavy earthquake, which made windows and doors shake. It gave everyone a fright, but as nothing further happened people went back to their usual business. Nothing happened on Monday, but on Tuesday and Wednesday mild quakes were felt again, and on Thursday there was a series of six really strong ones. Alarmed cables were sent to Cape Town and London, but the answers were reassuring. The volcano was dead, and a seismographic institute in South Africa had only registred some mild activity on Tuesday, which must have been quite far from Tristan. So there was no worry. But in the weeks to follow the earthquakes returned, and on Sunday September 17th there was a really bad one. People were in church and thought the roof would come down.

The Administrator had not been reassured at all by the messages from Cape Town and London, and sent observers to the other side of the island, and to Nightingale. This revealed that even the strongest earthquakes were not felt at all elsewhere. The conclusion was clear: the seismic activity was very localised, indicating that an eruption was imminent, exactly where the Settlement was located.

On October 8th, a large part of the cliff above the village came thundering down. Huge rocks tumbled accros the plain below, killing one sheep and, miraculously, damaging only one house. It was obvious that the Settlement was no longer a safe place, and the Administrator decided that the village had to be evacuated. The next day, suddenly the ground split. A sheep tumbled into the chasm, but reappeared when it was pushed upwards by the rising bottom of the cleft. The rising continued, forming a hill that grew higher and higher at a rate of about two metres per hour. The eruption had started, but no red-hot lava was yet seen.

The plan was to take everybody to Nightingale, with the two factory ships, the *Tristania* and the *Frances Repetto*, but that could not be achieved that day, so the people moved to their huts at the Potato Patches instead. The next day they were transported to Nightingale. Two ships that by coincidence were already on their way to Tristan and not far from the island, received the cry for help. The British frigate *Leopard* was coming with mail and cargo from Cape Town, and the Dutch liner *Tjisadane*, sailing from Rio de Janeiro to Cape Town, was scheduled to pick up a nurse and two girls at Tristan. The *Tjisadane* arrived first and took all the people on board. In Cape Town they were received cordially and were told they could stay as long as they wanted.

But London had different plans. The eruption came as a godsend. Ironically, one of the reasons the Colonial Office was in favour of evacuation was the cost of having an Administrator and his office on the island, something the islanders had never asked for. In practice, administration costs were largely covered by the sales of stamps and the tax income from the crayfish industry. Whatever, the volcanic eruption suited the Colonial Office very well. The people would be brought to England and stay there for the rest of their lives. Their opinion was never asked.

While the *Tjisadane* was still on the way with the refugees, the British ambassador flew from Pretoria to Cape Town, to make sure the islanders would not become too settled in South Africa, but would be transferred to Britain as soon as possible. On October 20th they sailed north with the *Stirling Castle*. In Britain, the idea was to disperse the Tristanians so that every family would find optimal opportunities to integrate into British society, but the islanders demanded to stay together. The authorities had to walk on eggshells, because the whole procedure was monitored by national and international press. So the dispersion plan was out of the question. There were several offers to house the whole population, for instance in the Shetland Islands, which might suit them better than the mainland.

In the end they were housed in Calshot, an empty complex of houses at an airforce base, not far from Southampton. Here, the Government really did everything they could to make life for the islanders comfortable, with social and medical support, and helping with finding jobs. And there were many private organisations assisting, like the Red Cross, and the Women's Voluntary Services (W.V.S.), who had cleaned all the houses and provided curtains and flowers. Sir Irving Gane, son of Douglas Gane, reopened the Tristan da Cunha Fund, and donations poured in from all over Britain and abroad. Medical support was essential, because the people were not used to British bacteria and viruses, and many of them were sick, in spite of the inoculations they got. Four islanders died of pneumonia during the first two months.

The Tristanians had trouble adjusting to the modern western world, where people would pass each other in the streets without a greeting. They were soon subjected to theft, and when an old man was molested by hoodlums, they decided that this was not their kind of world. Instead of integrating, they stuck together even closer. But in general, they had no trouble finding jobs, so the authorities thought that the transplantation was successful. It would never be safe to live on Tristan ever again, according to the Colonial Office. In Parliament the idea was aired to use Tristan for nuclear tests. In April 1962, six months after the evacuation, the jobs of the reverend and the Administrator were terminated. The Tristan files were closed.

But for the Tristanians the matter was not closed at all. After six months, they began to be supicious. All the answers to their questions had been evasive, and of course they saw what was written in the press, for instance about the proposal to use Tristan for nuclear tests. They wanted to leave. When Peter Munch, the Norwegian sociologist, visited his friends in Calshot, he saw faces and looks in their eyes which reminded him of his fellow prisoners in a German concentration camp.

In January the Royal Society organised an expedition to Tristan to look at the situation, taking two islanders as guides. The new volcano was practically dead. A large lava field had filled Falmouth Bay, and the beach had completely disappeared. So had the canning factory, but the Settlement was spared. Only one house was burnt down, probably hit by a volcanic bomb. All the cattle had vanished. Feral dogs got the blame, but earlier the *Leopard*, arriving after the *Tjisadane* had left, reported that all the dogs had been shot. Marauders had been on the island, removing all the sheep, geese and chickens. They even broke into the houses and took whatever they liked. Around the island a Russian whaler had killed most of the Southern Right Whales that used to live there.

The expedition had been asked not to come up with a report too soon, and certainly not to use the word repatriation. But unofficial stories soon leaked, and the two island guides came back with great optimism about returning to the island. The islanders were convinced the volcano had done no harm, and even saw it as a personal act of God that the houses were spared.

In April, as his last act, the Administrator informed the Tristanians about the findings of the Royal Society, but they only had one question: "When can we go back?" The Administrator said in vague terms that the Colonial Office had not made up its mind yet. The islanders were furious. They wrote a very angry letter to the Office with a copy to the press. The Colonial Office had only one strategy left: stalling in the hope that home sickness would eventually settle down and integration would follow naturally. They announced that a second expedition was needed, this time without island guides.

After two months, the Colonial Office still had not acknowledged the request for repatriation, so the Tristanians wrote a second letter. Not a request this time, but an ultimatum. If the government refused to arrange transport they would organise it all by themselves. The press loved it. Finally the authorities had to give in to the pressure and promised to co-operate with repatriation. The islanders started packing immediately. They would all be at home before Christmas. In August an advance party of twelve volunteers went to Tristan to make preparations for the return.

Still, the authorities kept dragging their feet, hoping the decision to go back was not as unanimous as the islanders wanted everybody to believe. Perhaps only the older folks had problems adapting. They could not imagine that the younger people would not see the benefits of living in a modern society. In December a representative of the Colonial Office came to Calshot to hold an anonymous vote. The islanders were dumbfounded. They did not need a vote, and had packed already. The hope was that in anonymity many young people would vote in favour of staying in Britain. Voters had to be 21 years old or older. Of the 153 voters, 148 voted for repatriation, 97 percent.

Being home for Christmas was not possible anymore. In March a group of fifty people were able to travel to Tristan, mostly family of the advance party, with the Dutch liner *Boissevain* via Rio de Janeiro. The others had to wait more than half a year. It was November 10th 1963 when everybody was back on the island, the whole episode lasting more than two years.

The problems were not over yet. To replace the lost sheep and poultry, animals were sent to the islands, but there were only two sheep per family. The Potato Patches had been ruined, and had to be re-planted, but the seed potatoes they got were of the wrong kind. For two years the harvests were minimal. So the islanders had hard times again as in the old days. And there were serious issues with the Administrator, who employed all the men for the construction of the harbour. However they often did not turn up, giving priority to their sheep and potatoes. The Administrator fired people at will, not giving them a new job for months, leaving them without income. People often survived on the savings they had accumulated during their stay in England, or simply fell back on their old subsistence economy. And there were incidents, such as a man sending his unemployed father to replace him for a day, thinking this was perfectly alright. But the Adminstrator sent the father back home and fired the son. And the Nightingale trips became a problem again. Cooking oil was now available in the island store, so the Administrator thought there was no need for any more of the fatting trips. He did all he could to end these trips. On the rare days that the weather looked fine enough to launch the longboats, and crews were already preparing things, he would call all the men to the harbour for work. It almost escalated into an open labour conflict. A woman working in the office had asked for leave to join her husband to Nightingale, and when this was denied, she packed up her things and quit her job. In the end the Administrator had to give in if he wanted people to work for him, so the Nightingale trips continued.

In general, the people were not as happy as they had expected to be, and they missed the luxuries they had grown accustomed to in England. Two years after their repatriation 35 islanders chose to return to England to find a better life. But back in Britain they were disappointed once more. There were problems with housing and finding jobs, and although the Colonial Office did all it could to help them, this time, they failed to feel at home again. In the end most of them returned to Tristan

In 1964 Peter Munch visited his friends again, this time in their own homes, and later, he saw them again in England. He wrote about his experiences in the book *Crisis in Utopia*, on which this chapter is mainly based. The islanders had lived through many crises over a period of 150 years, but this one was the worst. Many elderly

Europeans divide their personal history into 'before the war' and 'after the war'. The Tristan equivalent is 'before the volcano' and 'after the volcano'. They had learnt a lot about western civilisation. At least they now knew the meaning of the word bureaucracy.

22. A schoolboys' expedition

Mike Frazer had never heard of Inaccessible Island. He had just graduated as a biologist, and was in the pub contemplating his future. There he ended up in a converstion which must have gone like:

"Have you ever heard of Inaccessible Island?"

"No, I have not. Where is that?"

"Near Tristan da Cunha; it is one of the islands of the Tristan archipelago."

"Tristan? Where is that?"

"In the middle of the South Atlantic. It is teeming with seabirds. There will be an expedition going there, and they are still looking for an ornithologist. Would that be something for you?"

"When would I have to leave?"

"Wednesday."

In 1978, at Denstone College, a new headmaster was appointed. He had ambitious plans. He wanted to organise a real expedition, involving schoolboys, perhaps to the wilderness of northern Canada, or unexplored parts of Patagonia in southern South America, or maybe the interior of Mongolia. Michael Swales, in the position of Head of Science, had a different proposition. He wanted to go to Tristan da Cunha, in particular to Inaccessible, the least explored island in the archipelago. Michael's proposal was agreed, and he was assigned to organise the expedition, which he modelled after his own experiences on Gough Island 23 years earlier. The main goal was to produce a decent map, which, as was the case with Gough in earlier days, had never been made. Aerial photographs taken in 1961 failed to show details. It was a nice job for a team with a theodolite. And of course, as with the Gough Island expedition, there had to be a geologist, a botanist, an ornithologist and a zoologist for the other animals.

Just as with the Gough Island expedition, misfortune struck. First, the new headmaster, who was to have been the expedition leader and the ornithologist, had to drop out. No problem, Michael was well suited for both tasks. But, two weeks before departure, he fell ill and had to be replaced. Leadership was given to John Wooley, headmaster of a neighbouring school, who was also a proficient cook. Ornithologist Mike Frazer was added at the last minute. Three months later Michael also managed to get to Inaccessible where he joined the others. In total, thirteen people made up the expedition: seven adults with various tasks, and six schoolboys, between sixteen and nineteen years old. Three islanders were hired to join them as guides, Andrew Glass, Nelson Green, and Harold Green.

The expedition also needed a patron, who was found in His Royal Highness Richard, Duke of Gloucester, whose mother had laid the first stone for the Denstone Centenary Building in 1972. He was probably not aware of the link between his name and Tristan, where the ship *Duke of Gloucester* brought August Earle, and later delivered the Brides on the Beach.

In October 1982 the expedition members flew to Cape Town. Their equipment, including all the prefabicated parts of a large wooden hut, had alreday been sent by ship. They travelled from Cape Town to Tristan on the *Agulhas*. At first, the captain refused to take them on board because he found the whole operation totally irresponsible. Nothing was arranged for them to be picked up again, apart from a vague agreement with the navy. In all likelihood they would get stuck on the island for many months. But in the end he allowed them to board. After five days sailing they arrived at Gough Island with materials and a new team for the weather station. At dusk, thousands of seabirds wheeled around the ship, waiting for darkness before going to their colonies. At Tristan the three island guides came on board. Finally, on October 16th, the expedition reached Inaccessible Island.

Everything and everybody was brought ashore by helicopter on Blenden Hall Beach, on the exposed side of the island, but with the easiest access into the interior.

It took six days to build the hut, five for flattening a piece of ground to the right size and one for assembling the pieces. Then the daily routine began: early rising, an oatmeal breakfast, and then to work in small parties, weather permitting. And the weather was often not good. It was said to be the worst spring in twenty years. Usually it was possible to work only three out of every four days, and going to the high plateau was possible only once every four days. They stayed until February, including a month holiday on Tristan around Christmas, when they were picked up and returned by the islanders. A highlight for the island girls, all these handsome boys! There were a few fleeting romances, but no permanent relationships were established, and there was no 'hidden founding father'.

Food on Inaccessible was simple. All provisions were brought from the outside, canned or dried. But it did not take long before the island guides came up with fresh meat: petrels. There were millions, so taking a handful would do no harm. Below a petrel recipe, developed on Inaccessible:

Shearwater Marengo (for 16 persons).

Ingredients: 24 Shearwaters (assorted ages) cut into small pieces, vegetable cooking oil, powdered garlic, Cape Mixed Herbs, half a 7 lb. tin of peeled tomatoes with juice, 4 oz. concentrated dehydrated mushroom soup, 1 pint hot water, salt, pepper, 1 tin of butter, juice of two tired oranges, 4 oz. dehydrated sliced French beans.

First you fry the birds in oil, and then you use the oil to make a sauce with the other ingredients.

The birders charted all the albatross nests, and did some interesting observations on the endemic landbirds. There were the two buntings, the large-billed one and the small-billed one, and the thrush and the rail. The small-billed buntings flew around in two colour varieties, a brightly coloured yellow-orange one, and a drab, brown one. It had always been thought that the yellow-orange birds were adults and the brown ones juveniles. But now it appeared that both built nests, and the orange ones produced orange young, while the brown birds had brown offspring. Furthermore, the brown buntings nested in the tussock at sea level, and the orange ones higher up in the ferns. Could they be two species? Later, this problem was tackled by Peter Ryan, for his PhD thesis. With his wife, he spent six months in the Denstone hut. One of the three forms Ryan distinguished was named after Mike Frazer *Nesospiza acunhae frazeri*.

The thrushes were a nuisance. They always robbed the nests of the buntings after the boys found them. All the chicks that were marked with colour rings were killed by them. Not much new information was gathered about the diminutive rails, but great was the excitement when a nest was found. But those dreadful thrushes, always keeping a close eye on the boys, soon took the eggs.

The expedition left Inaccessible with the *Endurance*, which picked them up on February 10th, 1983, and took them to Tristan. From Tristan to Cape Town they travelled on the RMS *Aragonite*.

23. Moorhens rediscovered on Tristan

The last part of my masters in Biology was already dedicated to Tristan da Cunha. I made a zoogeographical analysis of the origin of the seabird fauna of the islands, using literature and museum specimens. When I visited the Natural History Museum in Leiden to look at old skins, the curator said to me:

"Ah, Tristan da Cunha, that is interesting. That is the island of the legendary Moorhen that never existed!"

What? How dare you! Of course it existed! I knew about all those people who ate them and found them so delicious! It was then and there that I decided to bring the Tristan Moorhen back and rehabilitate it. I dug into old sources to find eyewitness accounts and tried to trace specimens. To my surprise, I found a short note about birds collected on Tristan in the Zoological Museum of Cape Town, mentioning two Moorhens found between 1908 and 1910 by Mr Keytel, at a time when Island Cocks were already considered extinct. They were sent to me by Professor Winterbottom, in a package all the way from South Africa to the Netherlands. However, they turned out not to be Island Cocks. One was a Common Moorhen, the other one a Red-garted Coot, both probably vagrants from South America. The birds were sent back to Cape Town.

Precious little is left of the collected specimens, only three skins. One in the American Museum of Natural History (which I suspect to be a Gough Moorhen), and two in the British Museum. To my surprise I found a short note about Tristan Moorhen bones in the Museum of Cambridge University. I had to see those! So I went to England with three intentions. I was going to Tring to look at the skins, to Cambridge to look at the bones, and to Utoxeter to visit Michael Swales at Denstone College.

The bird collection of the British Museum in London had just recently been moved to Tring, to the northwest of London. Many skins were still unpacked. There should be two skins from Tristan, one in the collection of type specimens, the other one in the special collection of extinct species. A type specimen is the one that has served for the description of the species, so there is never more that one. They can never be replaced so curators guard them very carefully. The type specimen of the Tristan moorhen was hermetically sealed in a plastic bag, and I was not allowed to take it out. So I could not measure anything. The second bird, in the collection of extinct birds, could not be found. Perhaps it was still in one of the unpacked boxes. There was a third skin, but that one I found suspicious. On the label was written "Moorhen from the Cape Colony". A more recent label read: "This is not the Cape Coot, but probably comes from Tristan da Cunha". Later, someone had added in pencil: "yes". So this one could have come from either Tristan or Gough. I was soon done with the skins in the British Museum.

I spent a long evening with Michael Swales, listening to all his adventures on Gough Island and Tristan. He showed me hundreds of slides, even including some of Moorhens. But most of these were rather blurred, the birds being too quick all the time. They looked like brown streaks against an intensely green brackground. He did not mention the birds he released on Tristan.

In Cambridge I stayed with Mr Benson, a friendly, white-haired, deaf old man, who should have retired a long time ago, but was still taking care of the bones and skins in the museum. This is also where I got hold of a copy of the catalogue of Bullock's auction sale in 1819. Most of the skeletons were dubious, and looking at ambiguous notes on the labels, could either come from Tristan, or from Gough. Three turned out to be real, originating from the birds Sclater had used for his description of the species. There were also a few skeletons which definitely came from Gough. They where all without skulls because in preparing a skin the skull remains inside. I carefully measured everything I could, and perhaps I found a small difference in size, the Tristan birds being of a slightly lighter build than those from Gough, but don't ask a statistician. I published my findings in the *Bulletin of the British Ornithological Club* (Beintema 1972), edited by Mr Benson. As a budding scientist, this was my very first scientific publication in an international journal.

At the International Ornithological Congress in Berlin, 1974, I met Sir Hugh Elliott, Tristan's first Administrator, with whom I had corresponded about Moorhens earlier. He had seen my paper in the *Bulletin of the BOC*, and he told me he had a surprise for me. He had just heard that live Moorhens had been found on Tristan, coincidentally almost in the same year my paper had appeared, in 1972.

From October 1972 till November 1974, Mike Richardson had been Medical Officer on Tristan. Mike was not only a doctor, but also a keen birder. He climbed to the Base more than seventy times to collect data on the bird life, and he went to Inaccessible, Nightingale, and Gough several times. When Mike arrived on the island, the people told him that in March of the same year, they had seen Moorhens in an isolated, hardly accessible valley called Longwood, on the other side of the island. Mike visited the area three times, in April and December 1973, and in February 1974. He collected five specimens, which are now in the British Museum.

The area where the birds were found covered about eight square kilometres, and Mike estimated there were about 200 Moorhens. It was difficult to see them, but they were clamorous, and he could hear them calling all around him. He could not find any nests, but a male collected in December had swollen testicles, indicating mating time. In February and April he saw young birds, so he concluded the birds had an extended breeding season. It is the same with the birds on Gough, which may breed at any time of the year.

He looked at the stomach contents of the birds he collected. He found vegetable matter, including seeds, and grit and little stones, which birds in general use for 'chewing' their food after swallowing it. Furthermore, he found eggshell fragments and squid beaks, indicating that the birds would scavenge in seabird colonies. Spilled remains of squid are especially found around the nests of the Yellow-nosed Albatross. In the Longwood valley rats were very numerous, and Mike thought he saw prints of cats (which have never been seen since), so the conditions for the survival of Moorhens did not look at all promising. How was it possible that these birds were doing so well while the old ones were driven to extinction? What was their origin? I thought they very well could be survivors of the original stock which, through severe selection, had learned to run away from rats or cats.

The alternative explanation was that these Moorhens were descendants of the birds Michael Swales had brought to Tristan in 1956. In a correspondence with the Percy Fitzpatrick Institute in Cape Town he revealed that he had released eight birds not far east of the Settlement. I wondered how likely it was that a handful of Moorhens would have walked all the way across the mountain to the other side of the island, and, being from Gough, and therefore not adapted to predators, survived and multiplied, where the old ones could not. And if they lived there unnoticed for almost twenty years, they may as well have been there unnoticed for eighty years, as the area is seldom visited. So the identity of the birds Mike saw remained unclear. My favourite theory was that they were descendants of the original stock who had learned to deal with the rats. Michael said it was his successful reintroduction.

Part 2. To Tristan at last

24. Competition

In the austral summer of 1988-1989 I spent six weeks on Elephant Island in the Antarctic, with a Dutch colleague and six Brazilians, studying penguins. We lived in a small, container-like summer station, where we heard Shackleton's ghost walk over the roof at night. We flew from Pelotas, in southern Brazil, in a Hercules plane to King George Island, and then sailed with the ship *Barão de Tefe* to Elephant Island. In 1990-1991 I was back, this time with a larger Dutch expedition, hosted by the Polish in their base Arctowski (pronounce 'Arstofski') on King George Island. We flew in from Punta Arenas in southern Chile, and on the way back we were taken to Ushuaia in Tierra del Fuego on the Polish ship *Arctowski*.

Crossing the Drake Passage, there were large numbers of seabirds around the ship. Numerous petrel species, closely allied to the birds of Tristan da Cunha, and no less than six species of albatross. Being immersed for two southern summers in seals, penguins, and albatrosses, all my old Tristan feelings came back, and it was then that I decided it was time to go to Tristan. In addition, over the last few decades, DNA-analysis techniques had greatly progressed. So this is what I was going to do: I would go to Tristan, collect blood samples from Island Cocks, and then I would solve the riddle of the Moorhens once and for all, finding out where they came from.

It is not easy to get to Tristan. Shipping connections are limited. In 1991 there were the two fishing vessels with very irregular schedules, each having a capacity of around six passengers, with very long waiting lists, always giving priority to government officials and islanders who needed medical treatment. It might take years for an outsider to climb high enough up the list to obtain a berth. Then there was the annual visit of the RMS *Saint Helena*, all the way from England, on her way to Cape Town (no longer sailing today), and the annual relief visit of the *Agulhas*, changing the team of the Gough Island weather station. In fact the *Agulhas* was the only ship that would allow you to stay on Tristan for a few weeks, with rather reliable arrival and departure dates. The *Agulhas* had room for about 40 passengers, but also could have long waiting lists. Before booking, you needed approval for your visit from the Administrator and the Island Council.

Before looking into travel I had to secure my permission. When we were preparing our expedition in 1970 we had to apply to the Colonial Office, which in the meantime had been renamed Foreign and Commonwealth Office. So in early 1991, after coming home from Antarctica, I wrote to them. I heard nothing for three months so I wrote again. This time I got an answer. They said my earlier letter must have gone astray in a reorganisation and changing personnel. They had forwarded my request to the Administrator, and they warned me that receiving a reply might take a while.

After six months I received a letter from the Administrator, He explained that Tristan was not a holiday resort, and there were no hotels, TV or telephone. To see Moorhens I had to go on the mountain, which is only allowed with a guide. There would be no guides available for at least a year, because the male workforce was employed full-time in the harbour project. Furthermore, the islanders did not like nosy outsiders coming to visit. Landing would only be possible with approval of the Island Council. He did not say if he had forwarded my request to them. The message was clear. Please stay away.

I immediately wrote back that my original plan had already been postponed twice for six months, and that I had no problem with waiting another six. So, I suggested, I could come to Tristan in the autumn of 1993 after the harbour project had finished. I also said I would be happy to hire a guide for all the days I would be there, and would he please, please, forward my request to the Island Council. I also wrote to Tristan Investments in Cape Town, who were responsible for the bookings. Their answer was friendly but vague. For the next six months my chances on the waiting lists were zero, and they pointed out again that I first needed permission from the Administrator and Island Council.

I also revived my old Tristan connections. I wrote to Martin Holdgate, who remembered our correspondence of twenty years earlier. He advised me to contact Michael Swales (which I planned to do anyway), who was still involved in Tristan matters. Michael wrote to me that he was preparing a visit to Tristan, to look at the Moorhen situation. They had multiplied considerably since their discovery in 1973, and the people were wondering if they needed to be culled, as a dangerous, harmful exotic species. And, since Michael was held responsible for their introduction, he was asked by the Island Council to come and have a look. Dangerous Moorhens? I didn't get it at all. I had no idea what this could be about. I proposed to Michael that we should combine and co-ordinate our activities, but he did not reply. Months later, he wrote to me that his preparations had gone well, and that he was scheduled to travel in the spring of 1993. He had beaten me to it! I had waited for twenty years and now I was six months late! Unbelievable!

I also wrote to John Cooper in Cape Town, a Tristan and Gough veteran. I asked if his institute could help me to get to Tristan, and explained the Moorhen problem, of which he was already aware, of course. He wrote to me that he and the British scientist Gary Nunn were going to solve the problem using DNA analysis. This was Gary's specialty, and they had already obtained blood samples from Gough. Now they only needed samples from Tristan. I wrote to Gary Nunn, and offered him my assistence. I would be happy, free of charge, to collect blood for him on Tristan. He wrote back that he did not need my help, and that he himself was preparing a visit to collect the samples himself, next spring, 1993. How is this possible! For decades, nobody cared about the stupid Moorhens, and now, all of a sudden, I was facing two competing parties. I could not believe it.

Gary did not go. He could not spare the time. Michael wrote to me that Gary and John had asked him to collect the blood. I almost felt like everybody in the whole wide world was doing their utmost to prevent my going to Tristan. So it was a great surprise when I received a friendly letter from the Administrator, telling me that the Island Council had approved of my plans and that I was welcome on Tristan da Cunha. He told me which steps I had to take to obtain passage and explained the Moorhen problem. According to the islanders, Island Cocks were threatening the Yellow-nosed Albatrosses, by stealing their eggs. They thought it might be necessary to cull the Moorhens, to save the albatrosses from extinction. I could not believe what I read. Moorhens threatening albatrosses! Imagine! It all sounded to me like typical island folklore.

In fact, my visit in October would combine nicely with Michael's visit in April. He could take his blood samples, but in April there are no albatross eggs, so he could not look at egg predation. In October, on the other hand, they have eggs, so during my stay I could look at nesting success, and see if the Moorhens played a role in egg losses. Michael and I agreed on the division of tasks, and he suggested I should come and visit him again during the summer, after he returned and before I would go. He also advised me to contact Mike Frazer in Cape Town, the ornithologist of the Denstone Expedition. Tristan Investments let me know that I was on the list for the October sailing of the *Agulhas*, but I had to await the decision of the Administrator which of the people on the list would get a place on board. He would decide in July. I had my permission, at last, but it was not at all sure that I could go. I had to be patient.

In February 1993 Michael Swales travelled to Tristan for the third time. In the ten years since the Denstone expedition he had maintained regular contact with the islanders. He had also arranged that young people from Tristan could come to Denstone College for a year to study, so on the island he could be the guest of his own ex-pupils. Michael too had his logistic problems. In 1988 he wanted to organise a follow-up of the 1982-1983 Denstone expedition. In 1989 a team of six went to Fair Isle and the Shetland Islands for field training, but the expedition had to be cancelled when no places on board the ships could be obtained. In 1991 a new team was assembled, going to Fair Isle for training. But again they failed to get a passage so the team had to be disbanded. Finally, six places were obtained on board the fishing ship *Hekla*, sailing in March 1993. A new team was formed. But then came the message that departure was brought forward to February, and that instead of six, there would only be four places on board. It now fell in the middle of a school term, so taking schoolboys was out of the question. Selection was restricted to pensioners and unemployed people. The pensioners were Michael Swales himself and John Wooley, the cook of the Denstone expedition. The unemployed were two young people who had just graduated: Sara Wright, a biologist, leader of the training expedition to Fair Isle in 1992, and Chris Whirledge, a geodesist with expedition experience.

The return voyage was also not without problems. The idea was to return after three weeks with one of the fishing vessels, but they got stuck for seven weeks instead. They never managed to get a place on board. They were saved in the end by the RMS *Saint Helena*, visiting in April. That visit also brought Albert Veldkamp to Tristan, ex-pupil of Jan Brander, living in the same street in Vlissingen.

During their seven weeks' stay, they could spend frustratingly little time in the field. They were at Sandy Point on the east coast for a week, and a few days at the Caves in the southwest. And they made several daytrips to the base. But most days they were confined to the Settlement. A fundamental problem on Tristan is the availability of guides. You are not allowed to go on the dangerous mountain without a guide. In lousy weather there is no point in going up, and on days with nice

155

weather there are no guides, because everybody is busy with sheep or potatoes. It was quite frustrating at times, but at least Michael had plenty of opportunity to tighten the bonds between Denstone College and the people in the Settlement.

Island Cock research mostly took place at Sandy Point, and during day trips to the Base, usually just east of the Settlement. They had divided the island into sectors, defined by the radially running gulches, and within each sector the birds were counted. Highest densities were found at Sandy Point, not far from where they were discovered in 1973. They had since spread in both directions in the fern bush zone, and now occupied about two thirds of the island. The area where they occurred had expanded by about 500 metres a year.

In order to catch the birds to take blood samples various techniques were tried, using different types of nets and traps, often with limited success. The best method was to use a dog. A well trained dog is able to catch Moorhens without killing them. Hertbert Glass owned a 'softmouthed' dog, Dandy, and one day he came down from the base with a sack containing six live Moorhens. The story of Island Cocks stealing albatross eggs was confirmed wherever Michael asked about it, but he failed to find an eyewitness, who had actully seen it himself. The story gets even more bizarre. Michael was told that the nasty birds cleverly operate in pairs. One of them shows conspicuous behaviour in front of the albatross, which will then, out of curiosity, bend over to have a better look. Then the partner will sneak in from behind, punch a hole in the egg, and suck out the contents. That is how they do it. They also descend from the mountain at night to secretly steal chicken eggs in the Settlement.

There is an interesting notion that in the 19th century Island Cocks already had an equally bad, or even worse, reputation. When Captain Nolloth visited the island in 1856, he noted that the wild goats, which used to be very numerous, had become very scarce. He did not think they were exterminated by the islanders. He wrote about this:

"Wild cats, which must thrive on the numerous mice by which house and country are overrun, have probably, by destruction of the young, thinned the number of the goats, with perhaps the assistance of the predaceous island cock, which seizes chickens and attacks young lambs in the plain."

Moorhens certainly scavenge. Even our Common Moorhens in Europe do it. And when people see a scavenger eating from a carcass, they easily assume the animal has been killed by it. It is the same with vultures all over the world. People on Tristan say Skuas kill sheep.

A few years after Nolloth's visit, the goats were all gone. In the course of about two years they mysteriously vanished from the earth. Maybe they finally succumbed to inbreeding, being derived from only a few ancestors, or maybe they were felled by a disease. During those two years, sheep infected with scab had come to the island. Scab is not lethal to goats, but perhaps the already weakened animals could no longer cope with winters. Or were they all eaten by the ferocious Island Cocks... A few decades later, the wild cats also disappeared, equally mysteriously, so perhaps they were killed by Island Cocks too (also see chapter *Theories*).

25. Agulhas

In the summer of 1993 I went to England again. I wanted to have a look at the skins in the Britih Museum once more, and I had an appointment with Michael Swales, to hear about his experiences, and discuss my options. I also wanted to visit Ian Mathieson' second-hand bookshop *Miles Apart*, which specialised in the South Atlantic, and I was going to see retired reverend Patrick Helyer, in his house called 'Tristan'.

In the British Museum all the Moorhens were found without a problem, including the one they could not find in 1972, in the collection 'Extinct birds'. Also, I was allowed to take the type specimen out of its plastic bag to take a picture. And of course there were many skins from Gough. The five birds Mike Richardson had collected on Tristan in 1974 were deposited in the drawer with the Gough moorhens, on the assumption that that was where they belonged. But I could not detect any difference with the 'true' specimens from Tristan, so I kept the option open that Richardson's birds were survivers from the original stock on Tristan.

In the Moorhen drawer I found a letter from Mike Richardson to Sir Hugh Elliot, telling about his discovery. At that time, not many people knew about the birds Michael Swales had released on Tristan. But in his letter, Mike mentioned two birds that were kept in captivity by the manager of the lobster factory and had escaped. They were seen sometimes around the Settllement, but eventually disappeared.

Later I would learn that the situation was even more complicated. One of the captains of the fishing ships sometimes illegally traded in animals which he caught at Gough and sold in Cape Town. Sometimes when he was sailing around Tristan he received a message that he could expect an inspection. He would then just dump his collection overboard, often near Sandy Point. No problem for the seals and the albatrosses, and perhaps Moorhens would be able to reach the beach alive. So it is possible that apart from Michael Swales' birds, there have been more introductions. I now started to doubt my own theories and did not know what to believe anymore. We really needed the DNA analysis.

With the two museum specimens in my hands, as I stared out of the window a strange feeling of happiness overwhelmed me. In a few weeks time I would sit on a cliff with live ones in my hands, looking out over the endless Atlantic Ocean.

I discussed the problem of the availability of guides with Michael Swales. A point is that from anywhere on Tristan you can be back in the Settlement in a day, and islanders prefer to be at home. They are really helpful and want to be at your disposal, but never today because there is always another urgent matter to take care of. But tomorrow it is another today. This problem does not exist when you take guides to Inaccessible, Nightingale or Gough. So perhaps my best strategy would be not to rely on day trips, but to hire a guide for the entire period of my stay, and spend my time away from the Settlement. Perhaps it would be best to stay the whole time at Sandy Point, and just spend a few days in the Settlement at the end for sight-seeing.

Michael was secretary and treasurer of the Tristan da Cunha Association. Allan Crawford was still the president. Michael made me a life member. Most members were British of course, but I became the fourth from the Netherlands. Eighteen members lived on Tristan. The Association publishes a newsletter twice a year. Usually they are sent to Tristan by mail, but this time I could take eighteen newsletters to the island, and deliver them in eighteen houses, giving me a perfect opportunity to make contact with the islanders. Michael gave me a tape with a recording of Moorhen calls which he had recorded on Gough. I might be able to use these to lure birds into my nets or traps.

In July 1993 I finally got the confirmation that I had a place on board the *Agulhas*. And the departure date had been brought forward to September 23rd. Now I could book my flights, but I was late to do so, almost too late. I could only find one empty seat, five days before the ship was to depart. Having a few extra days in Cape Town was not so bad. I was staying with John Cooper, and he would introduce me to other Tristan veterans. We had a braai (BBQ) in Peter Ryan's backyard. He studied the buntings on Inaccessible. I also met Mike Frazer, who got a job in Cape Town after coming back from the Denstone expedition. With his wife Liz he made books about Cape nature. He wrote, she painted the illustrations. We looked at hundreds of slides from Tristan, Inaccessible and Gough, and we went bird watching around Cape Town. At John's institute I found some papers I had not yet seen, and at his home I saw Tristan books I did know about and he sold some to me.

Not many ports in the world have such a scenic background as Cape Town, with Table Mountain just behind the city. Like Tristan it is often covered with cloud (the 'Table-cloth'). From Table Mountain you can see the entire harbour complex. Slightly offside, I saw a bright red ship moored, the *Agulhas*. Red is the colour of South Pole vessels. It makes them easier to find in the white world in case of trouble. During the summer months, Agulhas operates in the Antarctic for research and logistic support for the South African base SANA. In autumn it goes to the southern Indian Ocean, to service Marion and Prince Edward, and in the spring (September-October) it sails to Gough, Tristan, and the South Sandwich Islands. It goes first to Tristan to deliver passengers and cargo, then to Gough to deposit the new crew for the weather station, and then drops a chain of weather buoys near the South Sandwich Islands, which will drift east in the course of the year. Then the *Agulhas* returns to Gough to pick up the old team, who have been training the new boys for two weeks, and finally back to Tristan to pick up cargo and passengers.

It was busy in the harbour with passengers arriving. Many of them were Tristan islanders. It looked as if half Cape Town was here to say goodbye to them. On the quay there were bundles of wooden beams, with name labels like Repetto, Lavarello and Green. Wood for building is a scarce commodity on Tristan, and waiting for a wreck has gone out of fashion. There was also a huge pile of beer crates. A few motor bikes were hoisted on deck, and even a car, labelled Repetto. It belonged to Gerald Repetto, who had lived in Australia, and who returned to Tristan as a pensioner, with his wife Rosemary, and son Cameron. Why would he need a car on Tristan?

Mike Frazer was also joining. He was going to Gough Island as Environmental Officer, to check if the South African staff at the weather station kept to the rules, and to educate them about flora and fauna. And he had to look for introduced plant species near the base. Alien species are often found around these stations, their seeds being brought there inadvertedly with building materials, clothes, and dirty shoes. Every year a new team of meteorologists and technicians comes to Gough. Nice boys, trained and selected for psychological stability to survive a year in isolation, but not necessarily aware of environmental issues. Therefore, the British government always sends an Environmental Officer to supervise the take-over.

Mike was greeted enthusiatically by a Tristanian woman, with a big hug and kisses on both cheeks. This was Joyce Hagan, mother of four daughters. Two of them will remember the Christmas holiday of the Denstone boys as the most romantic episode in their lives. Joyce travelled with three of them while the fourth waited for them at home. Her eldest daughter was Rosemary, Gerald Repetto's wife. The second was Marion, who was coming to Tristan with her future husband, Richard Collins, and a

161

baby girl. Richard was looking forward to his Tristan wedding and wanted to learn Tristan style masonry. The third daughter travelling with Joyce was her youngest, Debbie, who had been to Cape Town for a minor medical treatment.

Then there was Millie Repetto, who stayed in England after the volcano and got married there. Most of her children had grown up and left home. Now she was going to visit family on Tristan for the first time with her husband and youngest son. Evelyn Glass, who had worked as a nurse in Britain for many years, was also visiting family.

Conrad Glass returned to Tristan with his wife Sharon, and their little boy Leon, after having spent two years in England. Conrad, usually called Connie, was Tristan's police officer, and came to England for training. During his absence his cousin Clive Glass took care of things on the island as Connie's deputy. Connie learned how to chase fugitives on the highway in a Porsche, blue lights flashing. And he did a course in riot control. All very useful on Tristan.

Mike and I were not the only non-islanders on board. Of course there was the Gough Island team. Some of them went for a second term, and there was even one who was going to spend a year on Gough for the third time. All he could talk about was how many girls he would go after, back in Cape Town.

Brian and Heather Fredericks were going to inspect the school on Tristan. They had been involved in arranging for Tristan children to be sent to English schools, like Denstone College, and they had been doing school inspections in Saint Helena. This was their first visit to Tristan. The islanders would later call them the 'Education People'.

Petrus and Sonja van Rensburg came from Pretoria. They were going to collect the bones of a rare Beaked Whale for the Transvaal Museum. It had been found washed up on Tristan's south coast, carrying an unborn baby. It was probably a Sheperd's Beaked Whale, a very rare species, mainly known from strandings. The first one ever seen alive was observed in 1985 near Tristan, and a second one minutes later. The third record ever was one seen near Gough in 2002. A fourth was seen south of Tasmania in 2004. In 2012 a group of a dozen was filmed south of Portland, Australia, and in 2016 two groups were seen near New Zealand. So when Petrus and Sonja came to Tristan this species had only been seen twice, both times near Tristan. Up to 2006 there were six strandings on Tristan. The one that had been washed up in 1993 was fresh when it was found, but unfortunately too much time had passed, and the foetus had already completely rotted away. Islanders had cut the animal in pieces and packed it in plastic bags, ready for helicopter transport. Petrus and Sonja were called the 'Whale People ' by the islanders.

Often the expat doctor on Tristan is British, but now they had hired a South African physician, Peter Sandell, who lived there with his family. His wife Janet had been to Cape Town with their two little daughters for a medical treatment her husband could not offer. Peter worked at Camogli Hospital with a team of Tristanian nurses. The little girls, in white lace dresses and with long blonde hair, looked like fairies. They climbed indiscriminately onto everybody's laps. They thoroughly enjoyed life on Tristan, even having their own sheep.

There is no dentist on Tristan. In the past dental decay was unknown, but that has changed with a more affluent life style. Once a year a dentist comes to the island to fill all the cavities over a three-week period. This year it was Charlotte Rademeyer from Cape Town, with her assistant Ferdrika Hans. The last member of the medical team was Liz Bell, optometrist, who came for the periodical check of the spectacles, and to clean Nelson Green's glass eye. The islanders called her the 'Eyedoctor'.

Then there were the 'Hashmere People'. Hashmere is Tristanian for asthma. Professor Noe Zamel and his assistant Pat McClean came from the Mount Sinai Hospital in Toronto, Canada. Noe had specialised in the genetic background of lung diseases. As an example, only a small percentage of the people who are infected with tuberculosis, actually get the disease. The susceptibility is genetically defined. The same is true for the link between smoking and lung cancer. Noe wanted to find the gene responsible for asthma. Therefore, he was looking for isolated human populations with a high incidence. Tristan would be his ideal field laboratory, to collect blood samples for DNA research. He also found asthmatic populations on Easter Island, Tasmania, and a small Chinese island with only one family of 120 people. So he called his project 'Four Corners of the World Project'. He certainly had a taste for organising interesting travel for his work.

It took Noe four years to get his Tristan project off the ground. Like me, he first had to go to the Island Council, and as with me, the first response was negative. The People of Tristan would not be guinea-pigs. They already had their experiences. During their stay in England after the volcano blood samples had been taken, but nobody ever told them anything about results. Ten years later another medical team came to Tristan to take more blood samples and again, they never heard anything about results. So they had enough of it, Basta. Noe did not give up. He re-wrote his proposal, better emphasising the possible benefits for the islanders, but the response was again negative. In Toronto, Noe made a video clip about their work on cystic fibrosis, a nasty lung disease with a genetic background, and how they had identified the responsible gene. He sent his third request to Tristan, with the video. Doctor Peter Sandell showed the film and organised meetings to discuss the matter. The main message was that perhaps no cure for the islanders could be found, but that they might play a crucial role in finding a cure for the world. That did the trick. The people changed their minds and agreed to co-operate. Everybody was personally asked for permission to take a blood sample, and all but five people said yes. Sandell started with the preparations. He took samples of skin scrapings, as tiny dead skin fragments are often the cause of an asthma attack, and he collected dust from every house. And now Noe and Pat would take the blood samples, assisted by Sandell and his wife, and possibly the ship's doctor of the *Agulhas*. All samples had to be taken in the last two days in order to get them on board as fresh as possible.

Noe had to sign a secrecy clause in case his DNA analyses showed family relationships that did not exactly match the family trees. That was nothing new for Noe. He knew from his research in Toronto that five to ten percent of the children were not fathered by their 'own' father, who were usually unaware of this. No percentages are known for Tristan but everybody realises that these things are possible. And, they concluded, that was nobody else's business. Noe had prepared a personal certificate for everybody, and he had made a special T-shirt. And he promised to boost the local economy by buying everything from the tourist shop and the post office. But we (the other travellers) threatened to kill him and dance on his grave if he did that at the beginning of our stay on the island.

There is still one passenger left to deal with, Roger Western, the radio ham. There are about half a million radio amateurs, or 'hams' as they call themselves. They all have a transmitter and a receiver, and the idea is to assemble as many wireless contacts with other hams as possible. It is comparable to collecting stamps, monitoring car licence plates, identifying aeroplanes, or birdwatching. The kick is to obtain that rare stamp, to see that rare plane or bird, or to receive that rare radio signal. Rare signals have to be produced by other hams. There is a huge difference between senders and receivers. In principle the receivers are the collectors, while the senders form a smaller group of enterprising people, always looking for a new, special location to transmit from. Every ham will start his career by collecting contacts in as many countries as possible. They have a list of about 350. There are rules defining 'extra countries'. If an island is more than 250 km away from the motherland it counts as a separate country. Tristan is a clear example.

As a ham, you need proof of your contacts. The moment you receive the code of a rare sender you send back your individual code. The sender confirms receipt by returning your code plus an extra three letter code. Then you send a card to the sender, who signs it and sends it back to you. Hams collect contact cards. I realise that I am describing the situation as it was in 1993. Rules and procedures may have changed, I don't know. It takes the average ham about twenty years to collect all countries of the world. Some give up then, and start to collect something else. But the true fanatics specialise, for instance by doing the whole world again, using a different frequency. Or by going back to morse code. Most operators nowadays just talk, but there is a group which sticks stubbornly to old-fashioned morse signals.

Roger was an enthusiastic ham from the English Midlands. For years he only collected contacts, but three years ago he started to look for rare places to send from. He first wanted to go to Pitcairn, but someone else beat him to that, making 150,000 contacts worldwide. Instead, Roger went to the Solomon Islands, east of New Guinea, where he made 15,000 hams happy in just one afternoon. Roger specialised in morse so he does not talk. He has to single out one signal from the thousands that burst out together and send back the return code. On average he needs twenty seconds per contact. Roger has been in radio contact with Tristan for years, through the various radio operators on the island. First that was Pat Patterson, then Andy Repetto, and finally Ian Lavarello. Andy and Roger have become true radio friends. So, when Roger launched his plan to come to Tristan, he only needed to beep his friend and his approval by the Island Council was guaranteed. Many hams had already been in contact with

Tristan because Andy is a ham himself. But during his time he only made spoken contacts. Roger was going to do it in morse, making it rare again. In radio jargon 'roger' means 'I understood', so we called Roger 'Roger Roger'.

Finally, there were the VIPs, who were housed one deck higher up than the ordinary people: the director of Tristan Investments, Cecil Dickason and his wife, for the annual inspection of the lobster plant, and Gervais Gervasse, with the unmistakable air of a diplomat. Gervasse was going to replace Administrator Philip Johnson for a period of three months. Johnson was going to join us on the return voyage, and come back on the RMS *Saint Helena* in January. Gervasse was often on the helicopter deck, where Mike and I pointed out the various species of seabirds to him. Mike made it his task to give him a brief ecological education.

The first days at sea were very quiet. The sun was shining and the sea was calm. The meteo boys were sitting all day on the helicopter deck, half naked, drinking beer and playing cards. Mike and I were also always there, because the most interesting birds were following us in our wake. First Cape Gannets and cormorants, but soon the first petrels appeared. As soon as we lost sight of the coast, the gannets, cormorants and gulls disappeared. Most petrels were White-chinned Petrels, the Shoemakers of the old sailors, which I had seen around Cape Horn. And there were Cape Petrels, Southern Fulmars, and albatrosses. Surprisingly, the most common albatross was the Shy Albatross, which nests on islands around New Zealand, but is often seen on the Cape fishing grounds, just as the Royals from New Zealand are always seen around Cape Horn. Albatrosses are not afraid of covering large distances. French scientists, using transmitters, established that the Wandering Albatrosses from the Crozet Islands in the Indian Ocean, easily make trips of 15,000 km to find food for their chicks. Wandering Albatrosses appeared on the second day. They may have been Tristan Albatrosses because I don't see the difference and, in 1993 they had not yet been recognised as a separate species (that happened in 1998, based on Gary Nunn's DNA research). On the third day the first Yellow-nosed Albatrosses appeared, the Molly's from Tristan.

The islanders I spoke to on board all knew the Island Cocks. They were all convinced they were the ones that Michael Swales had introduced, and they all knew the story about them taking albatross eggs. To my great surprise, Joyce Hagen opened her handbag and produced a photograph of her husband Donald, sitting between the ferns, holding a living Moorhen in his hands. A man with a brown, withered face, a black-and-white dog at his feet, faithfully looking up at him.

On day four the weather changed. It was overcast and the sea turned grey, showing whitecaps here and there. The wind came from the northwest and the sea was starting to get choppy. Worse than that was a long swell, coming from the south, remnants of a distant gale. The Agulhas had a modern stabilising system, where large quantities of water can be pumped very quickly between tanks on either side of the ship, from left to right and back. These are also used for wobbling on pupose when breaking ice, but now they had to counteract the rolling movements caused by the swell. That did not function optimally. Somehow, we ran into interference with the wave frequency. We could go on without any sideways movements at all for a while, and then suddenly make one or two heavy lurches, almost throwing us off our feet. From time to time we listed to 25 degrees, with the result that nothing stays upright, everything flies off the tables, and loose chairs go everywhere. Sometimes we would reach 30 degrees, and then it is almost impossible to stay on your feet. And then, all of a sudden, we made a 45 degree roll, which feels almost vertical. Mike and I were on the helicopter deck, normally several metres above sea level. We slid down, together with some of the meteo boys, towards the sea which now washed onto the helicopter deck. But by the time we had rolled down there, the ship was righting itself again and we just escaped a soaking. It was quite scary. One deck lower there was panic. The water had reached the ceiling, and a door accidentally swung open, so seawater entered the ship, sloshing around in the corridors, flooding all cabins. The captain decided to change course to the northwest, delaying our arrival at Tristan, so the swell would no longer hit us broadside, but diagonally, reducing the problem.

Listing 45 degrees for twelve seconds, really was a scary experience. With 60 degrees and sixteen seconds the ship would not have been able to right itself again. It would have rolled through and capsized. It appeared that this had just happened to a Russian freighter near Gough. There was a radio message on the bridge, asking us to look out for floating rubble from the ship. We saw nothing

Wednesday September 29th 1993, 6 A.M. The ship was hardly moving anymore. The island should be right in front of us, but in the grey dawn we only saw mist and drizzle. A nasty breeze was blowing from the southeast but we were in the lee of the island. The sea surface was calm, but a long, slow swell was still rolling. Yellow-nosed and Sooty Albatrosses were graciously sailing around the ship, their wingtips sometimes almost touching the water. Behind the ship, tiny Wilson's Storm Petrels from the Antarctic were fluttering around and small flocks of Greater Shearwaters glided past at high speed on their way to Nightingale. Suddenly the fog lifted. Right in front of us a huge wall loomed, immensely green, with black lava cliffs where the swell broke into fountains of foam. We only saw the right half of the island, the left half still shrouded in mist. And there was the gently sloping plateau, dotted with cosy looking white cottages. It just looked like a small village on the West coast of Ireland. Behind the village a forbidding black vertical wall rose up hundreds of metres into the clouds. During a brief moment the upper rim of the 700 m high cliff was visible, and we even got a glimpse of the peak. Black and barren, with patches of snow.

Suddenly we were in the sun. Above our heads, there was a large triangle of blue sky, in the lee of the mountain. To the left and to the right we were looking at dense fog banks, rising up from sea level. It was like dragging a clean stripe with your finger on the bottom of a plate filled with cream. Only this plate was upside down, the finger was stationary, and the cream was being pushed along by the wind. In the sunshine the inhabited part of the island suddenly looked friendly and welcoming and the ocean changed from grey into a beautiful deep blue. We could see the small harbour, where waves were breaking between the two piers. But every now and then there would be two or three rollers that would not break. A small launch, carefully waiting for such a lull, appeared between the piers, and approached us, sometimes almost completely disappearing between the waves. It came alongside in the lee of the ship. Men were calmly looking up at us, hands in their pockets, while their boat was being tossed around like crazy. Some jumped onto the rope ladder that was hanging from the side of the *Agulhas* and like monkeys, they clambered up. Soon family members were hugging and kissing, tears in their eyes. Some had not seen each other for more than twenty years. Ropes were thrown across, to fasten the launch to the ship. Twice, a rope snapped, with the sound of a gunshot, when the boat fell away between two waves. Our luggage went overboard on the crane, in a large net that was sometimes almost in the sea, and at other moments was nearly crushed between boat and ship. Suddenly the launch was lifted high up on a wave, and the net landed on the bottom with a loud bang, four men hanging onto it, to keep it there and to unfasten the hook. Then all four of them were hanging in the air when the boat fell away again. But all went well and our luggage was brought ashore, more or less safely. Later I found that my new small laptop had not survived. I would have to do my writing the old-fashioned

170

way: pen and paper.

The boat came back to pick up the passengers. We were hoisted down in the 'bird cage', a wooden cage that could hold eight people. Like the net with the luggage, the cage was swinging from left to right and up and down. When the last people went down, there was hardly space in the launch to accommodate the cage, and those already there had some scary moments when it threatened to crush them. Miraculously, nobody was hurt. In front of the harbour we circled around, waiting for a calm moment between the breakers. Glassy blue hills of water calmly glided along, and when they reached shallowness they would suddenly rise up, and break with a thundering noise. On the pier an old man was standing, hand above his eyes, studying the wave patterns in the distance. Suddenly, his other hand went up, and there we went, full speed between the piers. A wave, not breaking, overtook us and lifted the stern. Thus we surfed into the harbour, now in full reverse, to avoid crashing into the concretre wall on the other side. When we commented on their skills, the islanders shrugged their shoulders, and laconically said: "We is used to it."

It looked as if the entire population of Tristan da Cunha was on the quay waiting for us. Bags and little children were lifted to helping hands and we clambered up a steel ladder. It was chaos. More hugs, kisses and tears. And suddenly, I had firm soil under my feet, but it felt like the concrete was still moving up and down. I was on Tristan da Cunha, at last!

Tristão's men exploring the islands

Shipwreck of Jorge de Aguiar

The classic picture

The Settlement

Gallinula nesiotis in Sclater's description, 1861

Moorhen in Tristan fernbush

Molly with chick

Subtle eyeshadows

View down from Hottentot Gulch

The wedding of Iris and Martin Green

Freshly moulted rockhoppers

Rockhopper portrait

Knitted penguins for sale

Island Cock knitted by Joyce Hagan

Greater Shearwater

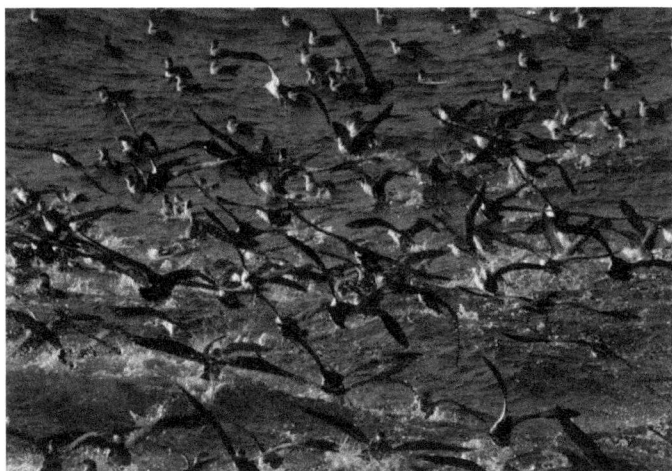
Shearwaters gathering in the evening

Brander's wooden shoes which I donated to the museum

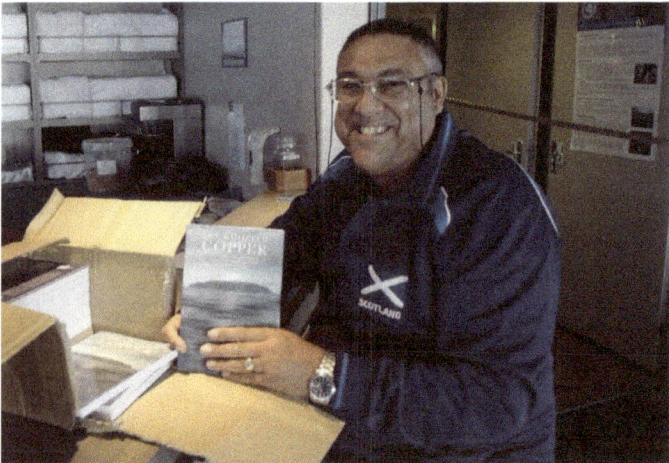

Conrad Glass selling his book *Rockhopper Copper*

Tristan Albatross

Fur seal pups at Gough Island

Nightingale Bunting

Wilkins' Bunting

Tristan Thrush

Inaccessible Island Flightless Rail (photo Troels Jacobsen)

26. The Settlement

Not everybody had to deal with the bird cage. Previously the helicopter left with the Whale People. They would fly directly to Deadman's Bay on the other side of the island, to pick up the packed whale. Then they would fly to Inaccessible Island to check the condition of the Denstone Hut at Blenden Hall Beach, and look at the small, unmanned weather station there. Mike was scheduled to join, to write an environmental report about the hut and its immediate surroundings. There was even the possibility that I could join, as a 'famous expert on flightless birds,' to see the sneaky rails of Inaccessible. John Cooper had already suggested this to the crew, and during the crossing Mike repeatedly stressed how important it would be to have me there. But when the helicopter suddenly departed without any warning Mike and I were still watching albatrosses. Mike was furious. Later it turned out that there was no place for Mike on the helicopter, and certainly not for me ("No joyriding," the Administrator said, with a stern look), because apart from the Whale People they also carried the acting Admin, and the crayfish director and his wife, plus luggage. VIP-treatment.

On the quay we were welcomed by the Administrator's wife. All foreign guests were assigned to an islander, who would show us where we were staying. The Whale People were staying with Herbert Glass and his family. Roger, of course, with his radio friend Andy. The acting Admin was lodged with the Administrator, Philip Johnson. The crayfish director had his own house next to the factory. All others went to the Quarter Deck, four houses available for government guests. Two were already occupied by technicians working at the harbour, the other two were for us. I stayed with Noe and the Education People, and the remaining four women (Pat, Charlotte, Ferdrika and Liz) were our neighbours. Each house had a living room, a kitchen, three bedrooms, and a bathroom. Gas came from two large bottles outside, water from the mountain, and electricity from the factory. The electricity would shut down at eleven pm for the entire village, except on fishing days, when the generators would run 24 hours to feed the freezing plant. Our houses were not the original Tristan dwellings, but modern wooden buildings. The only 'Tristan detail' was the door with two halves, like every other house on the island. Everybody leaves the door open all day, except for the lower half, to keep the rats out.

Our house was surrounded by a high hedge of New Zealand flax, imported for thatching when Tussock Grass was no longer available. But thatching is not carried out nowadays. The last thatched roof had been replaced by corrugated iron just one year ago. But the flax hedges also gave excellent shelter against strong winds. There were also various imported tree species, and around it all a stone wall with a white painted gate and flowering geraniums. It all looked very idyllic, but the total absence of birds was quite striking. The British guests in particular found this disturbing. These gardens needed Blackbirds, Wrens, and above all, Robins. The only songbird on Tristan is the Starchy, which stays away from people and is not found anywhere near the Settlement. The Education People seriously suggested imporing Robins.

The streets of Tristan are tarmac, but narrow and sometimes steep, only suitable for foot traffic with the exception of two 'through roads'. One runs from the harbour eastward, past the school to the new lava field with the rubbish tip around the corner. The other road branches off in the centre towards the west, past the Camogli Hospital, and then south to the Potato Patches, five kilometres away. Apart from cranes, bulldozers and other machines, motorised traffic consists of a handful of scooters and motor cycles, and seven cars (1993!), now augmented to eight with Gerald Repetto's car. In the village the speed limit is 25 miles per hour, and where the road passes the school there is a sign that hooting is forbidden. Anyone wishing to obtain a driver's licence has to go to policeman Connie Glass and do a test. Tristan licence plates all start with the letters TDC, followed by a number. Public transport is free. The bus, which can hold eight passengers, runs four times a day between the village and the Potato Patches. But, as with the ships to Tristan, your place on board is not guaranteed, priority always being given to the elderly.

On Main Street are the buildings of the Public Works Department and the supermarket, which is run by the Green family (Dutch grocer's blood always shows). A wide variety of items is on sale: boots, sneakers, fishing gear, birthday cards, torches, sewing materials, sugar, tea, coffee, flour, and shelves filled with cans, lots of cans. Better not look at the underside to check the expiry date. There is a small amount of meat and chicken for sale and, when a ship has been, sometimes large quantities of apples from the Cape. There are always potatoes.

Choice in clothing is limited, most of it being work garments. Yet, most women look nicely dressed in the latest (or one but latest) fashion, receiving their clothes by mail order from the outside world. Brightly coloured leggings are very popular (1993), tightly fitting sometimes slightly oversized legs. The older women often still wear loose, flowered dresses far below the knee, hand-knitted vests, and a scarf, just like in the pictures of the 1940s.

In the booze department there are cans of beer, sweet sherry and Cape wine in gallon cartons. Sweet sherry is especially favoured by the women. Bread is not for sale. There is no bakery. Everybody bakes their own bread. In the Quarterdeck we got our daily white loaf from Sandra Rogers, who also kept our kitchen clean.

"I comes in the morning," she said.

Between Main Street and the harbour is the radio building, where you can make phone calls through Radio Cape Town, every evening at six. Thus, people keep contact with family in South Africa or Britain and there is sometimes a queue. On the western side of the Settlement, above the harbour, is the Admin Building, where the

Administrator works, with his secretary. It has a nice hall where the Island Council meets, parquet floor, a picture of the Queen on the wall. And there is the immigration office, staffed by two part-time women, who enthusiatically put on their spotless uniforms when a ship turns up. They put a nice stamp in your passport which is valid for three days. If you stay longer you need an extension stamp. It took me almost two weeks before I found out that I had to show my passport at all, when Noe proudly showed me his passport with the beautiful stamp, showing the island and a passing albatross. The girls were not at all angry. They just assumed I would show up some day, which I did.

Shortly before my arrival the Admin Building had been equiped with satellite phone and fax, only to be used for Government business. For ordinary people it would be too expensive. The connection via Radio Cape Town was much cheaper. For me it was very convenient that I could do my last communications with the Admin by fax. Although I must say that a letter which takes months to arrive also has its charm.

Under the Admin Building is the prison cell which is also used for storing things. There is a story that the prison has been used once, when a drunken man had misbehaved towards a lady and used abusive language. He was sentenced to three days in the cell, but he could keep the key of his door himself, so he could go home for dinner. Allegedly, the main punishment was that everybody could see him making this daily humiliating walk.

There is also the video library. So it is not true that there is no radio and no TV on Tristan, as the Admin told me in his first letter. On radio, they can listen to the world news, and many houses have a TV set and a video player. There are no programmes on TV (1993) because there is no TV-antenna, but videos are very popular. Violence and soaps, confirming that life on Tristan is the best. Tristan has its own radio station. We accidentally tuned into it, when Noe was trying to find a Brazilian station.

"You are listening to Tristan radio," we heard, after a

little music and a few cooking tips.

"This is the eleven o'clock news," the voice continued.

"There is no news today."

Music. Knitting tips.

In the southwest corner of the Settlement, on the road to the Potato Patches, is the Camogli Hospital, named after the hometown of Repetto and Lavarello. Here, Doctor Peter has his consulting hours. Usually he does not have much to do as the islanders are seldom sick. When they have something serious they have to go to Cape Town with the *Hekla* or the *Tristania II*. But during our stay it was very busy at the hospital. Noe was there every day to let people do blowing tests, Charlotte was filling cavities ten hours per day, and Liz had installed her optometry equipment, which had survived the landing, thank God.

In the northeast corner of the village is the school, a modern, low building around a square patio. There are two classrooms, a workshop, and a kitchen, where the children have cooking lessons. When I went to have a look, girls were frying Tristan crisps in a deep pan filled with vegetable oil. Boys were sitting at their tables working on island projects. Children go to school until the age of fifteen. If they want to learn more, they have to leave the island for additional education in Cape Town, Saint Helena, or Britain. Teachers are no longer expats. Teaching is done by Tristanian women nowadays. The school also hosts the library, which is well stocked. I estimated the total length of filled shelves to be 250 metres. The various categories are indicated with felt-pen written labels, but often the titles do not fit. The only category that seems to be kept up carefully is 'sea stories'. There are no true Tristanbooks in the public library. They are rare and in great demand. There are two sets on the island: one in a locked cabin in the teachers' room, and one in the Admin's office. Sharon Glass, the policeman's wife, is the librarian. There is a small notebook on the table, in which you write your name, the date, and the title of the book you take. When I went there on a rainy day, the previous page only had the names of my travel companions. Reading is not a popular pasttime on Tristan. Most reading is done by the children, but I guess the same is true for my hometown.

Close to the school, the famous longboats are lying along the road, in a neat row, upside down, secured by ropes. The longboats are not used for fishing. They are only used for the traditional trips to Nightingale, to collect guano, firewood, eggs and petrel chicks. Going for guano and firewood no longer occurs, artificial fertilizers being much better, and gas replacing firewood. But the egging trip and the fatting trip still take place. These are also not really necessary anymore, vegetable oil from the supermarket replacing petrel fat. But allegedly, nothing beats potato cakes fried in petrel fat. The longboat trips to Nightingale are also holiday outings just for the fun of it. Every longboat has its own crew, and although they have to stay together for safety, it always ends as a race. The rigging is rather primitive and does not allow tacking against the wind. So they have to wait for a nice easterly wind to make the trip and stay there until a westerly wind can carry them home again.

The post office is a brown building in the style of the Quarterdeck. Two people work full time to send stamped covers to collectors all over the world. Ian Lavarello spends hours stamping stamps, before he goes to the radio station at six o'clock for the daily connection with Cape Town. Like Connie Glass, Ian spent two years in Britain, for technical education. He was important to me because he was one of the two ornithologists on the island, the other one being Connie Glass, with whom I had spent a lot of time on the helicopter deck of the *Agulhas*, bird watching.

In the centre of the village is St. Mary's Church, which needed three generations of nagging reverends to succeed in having it built. The churchbell is not in the tower but hanging next to the door. It was the ship's bell of the *Mabel Clark*, which was wrecked in 1878. The catholics, established when the Smith sisters came to Tristan in 1908, have their own little church, not much different from an ordinary house (a real church has been built for them in 1998).

The largest building is the Prince Philip Hall, the community hall. Every Saturday there is a dance. There is a lot of traditional Tristan music but much of that has been forgotten. Traditional dances, like the famous 'pillow dance' are only performed on special occasions. Most dancing nowadays is disco style with pop music. Sometimes the local band 'Tristan Rock Lobsters' perfoms, singing songs like 'I's born in Triss'n'.

The local pub, the Albatross Bar, is also in the Prince Philip Hall. The pub maintains strict opening hours, closing at nine pm. During the Saturday dances, people have a solution for that. At ten pm they have the 'Tea Break'. Everybody goes home, or joins a friend, to have 'tea' (usually something stronger), before returning to the dance. The bar is closed on Sundays, but fortunately the cafe, which also serves beer, is just opposite, and open on Sundays. It reminds me a little bit of old stories of the Irish west-coast, where after closing hour pubs would only serve guests from elsewhere, resulting in entire villages swapping.

The typical Tristan house is small and very low to allow the gales to pass over. The gables are hewn out of soft volcanic rock. From the ridge-pole (formerly from a ship's wreck, nowadays imported wood) the roof at the back almost reaches to ground level. At the front it ends high enough to accomodate a window and a door, but you often have to bend to pass. With high humidity, and firewood being scarce, houses tended to be moist and mouldy in the past with their porous walls. People used old newspapers for wall paper. In modern houses walls and ceilings are covered with masonite, with glossy paint in colours like blue, pink, or yellow. One house has bigger rooms and larger windows: the Residency, where the Admin lives. A neatly trimmed lawn and well kept flowers, British flag hoisted, and a magnificent ocean view. On our first night ashore, we were invited for a reception. I appeared to be the only one who never thought of bringing a tie to this island. We had a drink and toast with fresh Tristan Rock Lobster.

Next to the Residency is the museum annexe craftshop, managed by Catherine Glass. The museum only opens on Wednesdays except when there is a ship visiting. Then it is open every day. There are two parts: the natural history, and the history department. There is a large collection of rocks, including the rare mineral tristanite, and there are pinned insects, like the endemic flightless moths. There is also a 15 cm large moth which was blown to the island all the way from the Brazilian rain forest. The history department displays all sorts of things left on the island after shipwrecks and old photographs. There were no wooden shoes from the hundreds that Jan Brander sent to Tristan in 1937.

In the craftshop you can buy small scale models of the ox-hide moccasins people used to wear in the past, and models of the longboats. Typical Tristan coasters appeared to come from Hong Kong. The most special are the knitted penguins. As far as I know, Tristan da Cunha is the only place in the whole wide world where you can buy knitted penguins. Knitting is quite a thing in Tristan. Women are said to always knit, even when they walk the five kilometres to the Potato Patches. Most famous are the Tristan socks, long white socks which you are supposed to wear over your trousers. They have one, two, or three coloured bands close to the top, often blue. In the old days girls would knit socks for a potential boyfriend, the number, width, and colour of the stripes indicating how much she loved him.

Next to the school is the cemetery. In a separate walled part, are the graves of the first settlers. William Glass's grave has a white marble stone which his sons had sent from America. On most other graves inscriptions have withered away in the soft volcanic stones marking them.

Nobody could tell me which grave belonged to Pieter Groen (in 1996, three years after my visit, Dutch radio reporter Marnix Koolhaas visited Tristan. Using old photographs, he managed to identify Groen's grave).

Walking around in The Settlement, you can always enter any of the houses. It is still the custom that nobody locks his door and, if the weather is not too bad, everybody leaves the door open (or the upper half), all day. I often dropped by at Joyce Hagan's house, also because Richard and Marion were staying there. Joyce makes excellent potato cakes. The classic potato cakes follow the recipe originating from the time when flour was a rare commodity on Tristan, and the entire pastry business was based on potatoes. In the craft-shop I bought a booklet with recipes, published by the Mothers Union of St. Mary's Church, written by Pamela Lavarello. The basis for potato cakes, and many other baking products, is the potato mixture, made of one pound of mashed potatoes, five ounces of flour, and a pinch of salt. Roll a layer of half an inch, cut out the cakes with a knife, and fry them in oil till golden brown. Serve with sweet cream or jam. Joyce served them with orange marmalade.

As there were no ovens in the old days, things like apple pie, berry pie, or jam rolls were made by simmering them in boiling water, sometimes wrapped in a cloth, for one or two hours. Most recipes in Pam's cookbook are not difficult, but some ingredients may be hard to get if you are not living on Tristan. For instance if you need a large quantity of lobster tails for the soup.

27. Sandy Point

On arrival in the Harbour, after my adventurous crossing with the *Agulhas*, the Admin's wife introduced me to Jeff Rogers, who had been appointed to be my personal guide. He would be available for me every day of my three weeks' stay. That evening I was invited to have dinner at Jeff's house, and the next day he would take me to Sandy Point, where we would start marking Molly nests to monitor their hatching success. Following my latest correspondence with the Admin, he had organised an open tender. Four men had applied to become my guide and Jeff was chosen. It looked as though everything had been perfectly organised. I would not suffer from the kind of frustrations Michael Swales had to endure in April, getting stuck in the village for weeks on end.

We had dinner in Jeff's living room, all furniture against the walls, which were covered with masonite, painted glossy yellow. A large stereo tower was prominent. We filled our plates in the kitchen with potatoes, lamb, carrots, peas, and cabbage. A glass of sweet sherry as appetizer, and pastry as a dessert. Jeff confessed that he had invited me to get an impression of my physical condition. He asked me about walking and climbing experiences and was clearly inspecting me. The mountain is tough going and he was certainly not prepared to drag an old, unfit person along. I passed the test and he even decided that I looked fit enough to leave the radio at home, which was meant to give a daily report on our wellbeing. It weighed four kilograms so it would make a huge difference in our luggage to leave it here. He seemed to be looking forward to our expedition. I certainly was. We would go to Sandy Point for ten days. We would stay in the hut at Sandy Point which had been freshly painted especially for my visit. A large sack of potatoes and a container with four litres of vegetable oil for cooking were already there, so we only had to carry some supplementary food items. However, leaving the following day, Thursday, as originally planned, would not be possible. Too many things to do, including shopping for our trip. So I spent Thursday climbing the new volcano, still steaming, and the Hillpiece, to enjoy the scenery. It was a windy day with chilly, horizontal rain.

Friday, at last. Jeff came to fetch me at 8:15 am. His dog called Number, was dancing around his feet, enthusiatically wagging his tail, apparently very happy to join us. I was less happy. The dog would certainly diminish my chances to see Island Cocks. But Jeff explained that an Islander never leaves the Settlement without his dog. We divided the cans of food between our rucksacks. I still had a tax-free bottle of whisky from the ship and Jeff brought a bottle of sweet sherry. I had found a number of bamboo sticks to mark the nests but we decided to leave these. Jeff said there would be enough tree branches to use instead. It was a beautiful sunny day, not a whisper of wind. The sea was like a mirror, and the swells came from the southwest, so we could take the risk of walking all the way along the boulder beach instead of taking the difficult trail across the mountain.

Off we went. The original plan of going for ten days had started to erode. Jeff said we could not possibly miss the wedding on Saturday. His niece Iris Green was going to marry Martin Green, and a true Tristan wedding was something I certainly wanted to witness. I had received an invitation and it would be very rude not to turn up. So we would have to come back on Friday. OK. The next problem was that we could not easily take more food with us than we would need for five days, so we would have to come back on Wednesday. OK. Then Jeff said he actually wanted to be back on Tuesday, because that was his wife's birthday. I was more than welcome to come for a drink too. Tuesday. OK.

We walked through the village towards the east at full speed. Jeff was in a hurry, apparently. The *Agulhas* was still at anchor. Since arrival the sea had not been as calm as today so off-loading was in full swing. Heavy equipment was loaded onto large rafts, and then towed into the harbour. This included Repetto's car. The raft was almost invisible between the gentle swells so it just seemed as though the car was riding the sea which looked very strange.

On the other side of the new lava field, not far from the rubbish tip, the Whale People had established their 'kitchen'. Petrus was cutting brownish black chunks of rotten blubber and meat from the bones with a huge knife, and Sonja was stirring the soup in a large oil drum, cooking the last meat off the bones, over a wood fire. The stench was absolutely overpowering but the smelly steam drifted towards the east, away from The Settlement. The whale was surpsisingly small, not much bigger than a dolphin. Unfortunately the narrow pointed beak was severely damaged during transport.

"Are you going to Sandy Point?" Petrus asked.

"Yes," I said.

"When are you coming back?"Sonja asked.

"Tuesday," I said.

"Monday," Jeff said.

We ran further east, hopping from boulder to boulder, Number way ahead of us, than Jeff, way ahead of me, than me. Jeff was in a hurry, and I had trouble keeping up. Also I wanted to take a picture every now and then of a stray penguin, a seal, or the spectacular landscape. I was sweating profusely. I never thought it could be so warm on a sunny day on Tristan. This was supposed to be a subantarctic island, but it felt more like subtropical and the heat felt oppressive with the high humidity. Sometimes Jeff disappeared around the corner. After a while I would find him, sitting on a boulder with a cigarette in the corner of his mouth and looking displeased. As soon as he saw me he jumped up, not waiting for me, and ran away again. So we continued boulder hopping, the sea on our left, and to the right the enormous wall of rock, rising up and looming over us. Sometimes the wall was intersected by deep gulches. Most of these were narrow and steep with small waterfalls. The less steep parts were densely vegetated with trees and ferns in an impenetrable tangled mess.

Jeff was in a hurry because the wind was turning towards the east, and before arriving at Sandy Point we might get into trouble with the surf. In any case we had to pass two bottlenecks which he wanted to be beyond before he was prepared to make a stop. Here, the beach was so narrow that the surf reached the vertical wall, breaking into fountains of white foam, followed by the the hissing and rattling sound of the backwash and rolling pebbles. Waves alway come in groups, so if you wait a while, carefully studying the pattern, you may find a pause between the waves, allowing you to pass, just like waiting for the right moment to steer a boat into the harbour.

"If I says run, you runs and keeps runn'n," Jeff said, adopting a starting pose, Number between his knees attentively watching the sea.

"Run!" he suddenly yelled, slapped Number on the back to send him forward, and we ran after him over the wet boulders, concentrating on our footing.

"Go, go, go, go, go!" Jeff continued calling until we were on safe ground. Right behind us a large wave hit the wall, sending foam up in the air several metres. Near the second bottleneck we had to cross a rock protruding into the sea, three metres up, almost vertically. Jeff first, than me, leaving Number behind, yelping in panic. I may not be a fast runner but I am a good climber, so I was up there in no time, not needing Jeffs helping hand. Instead, I used my own free hand to help poor Number up.

"You take care of yourself, the dog will come with us anyway," Jeff said.

We passed another go-go-go-go-go cape, and then a kilometre clambering large rocks from a recent rock fall on a narrow beach between the sea and the cliff. Most of them were dry, but in places the surf would splash over them, reaching the wall.

"Never keeps your eye off the sea," Jeff warned, but I could not avoid getting a wet leg. Finally we arrived at the black sandy beach of Sandy Point. We had to clamber up a steep, fern-clad slope, and then we entered a patch of pine forest with an undergrowth of brambles. A little piece of Europe. And there was the hut, freshly painted, a sack of potatoes waiting for us at the door.

At Sandy Point the slope is less steep and it is only 300 m to reach the Base. The slope is densely vegetated with fern bush. There is a low, undulating plateau of a few hectares where the forest has been planted. This is the most sheltered location of the island. There is also a small grove with apple trees, and even a few peach trees.

"Cack, cack!" came from the bramble bush. Moorhens! Apart from that call there was total silence. A little wind through the treetops, the rushing sea down below, limitless views towards Cape Town. Not a singe sound related to human civilisation. Or... a deep rumbling sound approached, and suddenly there was the helicopter, on a sight-seeing tour, coming back from Inaccessible. They could have easily taken us and dropped us off here, which would have allowed me to make an estimate of the density of nesting albatrosses in the southeast quadrant of the island.

Mike had been lucky. When the helicopter deposited the VIP's on Tristan on the day of arrival, and had continued with the Whale People to Deadman's Bay to pick up the whale, the flight to Inaccessible had to be cancelled because of dense fog over there. Therefore, the flight was made today. Me joining was out of the question.

"No joyriding!" the Admin repeated severely, deaf to Mike's pleading how important it would be for me to see the rail. Instead, he joined himself, and invited the VIP's on a fun trip. Mike's luck was that there was no luggage so he could come and write his environmental report.

In the hut we made our beds. Our home was through and

through humid and clammy, smelling of mould and paint. The masonite-covered walls were painted glossy blue, Tristan style. My shirt was totally soaked. Overheated feet up for a while, a chunk of bread with sardines, and a swig of whisky.

After lunch we went to the penguin rookery. Jeff wanted to see if there were still some fresh eggs to be taken and I wanted to see the penguins, of course. Next to the forest, there is a small area of undulating grassland, where a few head of semi-feral cattle graze. They see humans so seldomly that the only way to secure one for meat is to shoot it. We followed a narrow cattle trail through the meadow, about thirty metres above sea level, along the edge of a vertical cliff, looking down on the beach. Silent Mollies glided past, and on the higher slopes behind us, several were sitting between the green ferns. That is where they have their nests. From the cliff, raw eerie cries rose up, from the Sooty Albatrosses, the Peeos. They nest on narrow rocky ledges, often way out of reach, but here was one sitting on its nest almost at arm's length. The shades of brown, almost black, and lighter grey, in the coloration of this bird, are unbelievably subtle. Pure beauty. In flight, they are among the most elegant birds on earth, especially when they perform their courtship flight, where two birds describe wide arcs in the sky in perfect synchrony.

The calls of Skuas warned us we were close to the penguins. Where there are Skuas, there is Skua food. Penguin products. We zigzagged down a steep, slippery fern slope, into a gulch leading to the beach. On the other side the huge rock wall reached the sea again, with a narrow boulder beach which ended a little further on. It is impossible to go further south from here. I was surprised to see that we were so far around the corner that we could see Nightingale in the distance, looking southwest. I had never seen it before but it was such a familiar silhouette.

There were three small penguin colonies, one on the beach, one in the mouth of the gulch, and one higher up in the gulch. Rockhoppers are closely allied to the orange crowned Macaroni Penguins I knew from Elephant Island. They are slightly smaller and their waving yellow head plumes are much longer. But they have the same mean red eyes and sharp red bills you had better stay away from. And like the Macaronis, they are true mountaineers, climbing and hopping up steep rocky slopes, dozens of metres above sea level, sometimes hiding their nests between the ferns. Penguins in green, very different from Antarctica.

The lowest colony was the most densely populated, with nests between the boulders, lined with grass and seaweed. Like the Macaronis, Rockhoppers first lay one egg, and then replace it with a second one, slightly larger than the first. The first egg rolls away and is wasted so they only raise one chick. It is totally incomprehensible how this kind of behaviour could have developed. In some colonies the discarded eggs all roll in the same direction and pile up. What a waste! Feast for the Skuas. Jeff paced through the colony, carefully avoiding the bills, looking for surplus eggs, which are delicious, so he said. Most loose eggs were broken and rotten. Jeff took the five best, but later these too appeared to be foul and inedible.

Rockhoppers nest in large numbers on Nightingale. They used to be very numerous on Tristan too, but they have been brought close to extinction and are no longer nesting anywhere near The Settlement. People not only took their eggs, but also killed them for bait for fishing. In addition they took their scalps, which were sown together to make the traditional decorative 'tasselmats', the yellow plumes radiating. Tasselmats were used in bartering with ships, as souvenirs in exchange for goods. Depending on the size, you needed six to fifteen birds for one mat. Today the penguins on Tristan are protected and are gradually increasing in numbers again. Only on Nightingale are people still allowed to take eggs. The impact on the population is negligible now that only the discarded eggs are taken.

We dined on sardines and Tristan crisps, fried in two litres of moderately heated oil. Outside, Moorhens were calling. I tried to lure them into my mist nets with the sound I had on tape but to no avail. A few obscure glimpses were all I got. At seven pm it was getting dark and, as there was no light in the hut, we just went to bed very, very early and slept soundly after an exhausting day.

Six in the morning, another gorgeous day. Bright sunshine, no wind, a cloudless blue sky, and an even deeper blue sea without any whitecaps. With a large bundle of eucalyptus sticks we had cut the previous day, and a can of blue paint we went uphill. Number ran ahead of us, tail-wagging, making sure we would never see a Moorhen. On the plateau two cows looked warily at us, and we met some grazing sheep. Above the grazing line we entered a dense vegetation of ferns on a ridge between two gulches. Within ten metres we found our first albatross nest hidden between tree ferns under an Island Tree. The bird clapped its bill but did not give any further response. Number was very excited but the bird took no notice of him.

"Number! Leave it!" Jeff yelled.

"Number! Get back!"

I heard this repeated many times during the day, because Number was young, playful, and disobedient. And indeed, we never saw a Moorhen. We heard them often enough, but they never showed themselves. Around Sandy Point they are very common.

Albatrosses build a small tower on which to nest, by scraping mud together. The nest is about thirty centimetres in diameter and about the same height. It is built in the centre of a bare spot of a square metre. On the top, they trample a neat cup, in which they lay their single egg. The Yellow-nosed Albatross is an unbelievably beautiful bird. Its plumage is of the purest white, with a subtle grey shade around the neck, and a stern-looking dark eyebrow patch, fading into the white cheeks. They look at you in silence, slowly waving their head from left to right, as if to show

200

you they have no idea why you are there. Sometimes they clap their bill, and if they are getting more excited, they contract their cheek feathers to show a bright yellow line running backwards from the corner of their mouth, making a high, thrilling, rattling sound. The bill is black, with a bright yellow line running along the top, from the tip of the bill to the forehead.

Yellow-nosed Albatrosses breed on Tristan, Inaccessible, Nightingale and Gough, and on Amsterdam island and Saint Paul in the Indian Ocean, but the majority of the world population nests on Tristan, about twenty thousand pairs (when I visited in 1993 the populations of the Atlantic and Indian Ocean had not yet been separated as different species). On Tristan, they nest on the Base, mostly on the eastern side of the island above Sandy Point and Stony Hill. On the western side they have been virtually wiped out, their chicks being considered a delicacy. A small colony is still present near the Settlement, above Hottentot Gulch. They are protected now and the birds in this area are being monitored. This is the so-called Molly Plot, where the two bird men of the island, Ian Iavarello and Conrad Glass look at nesting success, and where John Cooper wanted me to have a look.

Each nest we found was marked with a eucalyptus stick, the end dipped in blue paint, and with a number written on it with a black felt pen. Hatching success is calculated from the number of nests in the sample, and the length of the intervals between marking and checking. I had calculated that, with an interval of about two weeks, I would need a sample of at least sixty nests, so I optimistically aimed at one hundred. That was not going to be easy. At the end of the day we did not get any further than thirty-three. There were more than enough birds, but they were far apart, and often well hidden between the ferns, in very difficult terrain. Where the leaves of the tree-ferns formed a solid roof, about waist level, it was impossible to see where you put your feet, so you would stumble on old trunks, or step into a deep hole between them. In sheltered places the Island Trees could grow to more than three metres with ten cetimetre thick trunks, often growing in horizontal bundles hidden by the ferns, almost impossible to pass through. Bruised and scratched shins, and it was unbelievably hot, with a near tropical humidity. In open spaces grew smaller fern species and sometimes thick, soggy cushions of peat moss and beautiful clubmosses. And all around us were these serene white birds. It was magically beautiful.

When we were watching the ocean during a brief rest, we saw the *Agulhas* sail past on her way to Gough Island, dragging a long line through the blue ocean. We were sitting on a small grassy ledge above a deep ravine, where Mollies were showing off their aerial acrobatics. The would catch the updraft and hang motionless in the air, sometimes two close together, using their enormous feet as flaps, to fine-tune their position, necks curved in a strange way. They uttered their high rattling call, declaring their love to each other.

Number was running around, happily tail-wagging, making sure the Moorhens would keep their distance. On the grassy cliff edge he found some holes where he started to dig enthusiatically.

"Haglets," Jeff said. Atlantic or Great-winged Petrels.

"Number! Leave it!"

In contrast to most other petrel species Haglets breed in winter so at this time, they should have large chicks, almost ready to fledge. They are no longer being taken for food by the people and, after having been brought close to extinction (also helped by the rats), they are now fully protected.

On some ledges sat rows of albatrosses without nests. We also saw old, withered nest towers, overgrown with mosses and no longer in use.

"There's Mollies, but no nesses," Jeff said with a meaningful look in his eyes. He did not exactly say so, but his meaning was clear: Island Cocks were responsible.

While Jeff was preparing dinner (Tristan crisps and

sardines), I took the opportunity to have some moments in the forest on my own, without Number, and without Jeff calling Number back all the time. Near the penguins I climbed up in a gulch, through a green, mossy tunnel with giant boulders under a dense roof of Island Tree canopies. The absence of the dog was rewarded. In less than five minutes an Island Cock and I were watching each other at close range. If I could close him in against that big boulder I would just be able to grab him. After a silent moment he sneaked away in a small gully which looked like a dead end. This was my chance. I dived onto him, but it appeared that the gully had a small backdoor. "Chack Chack" he said triumphantly from behind a boulder. Three more times I managed to see the bird and I got the impression that he was just as curious as I was.

Sunday. Beautiful weather again, but slightly hazy. Another sweaty day through the green wilderness. Twenty-two new nests, bringing the total at fifty-five. That was all. On the way back to the hut we tried to hunt Island Cocks. Jeff had one cornered, which he could actually touch, but at the last second it managed to escape. Number was more successful. A good Tristan dog knows to catch Island Cocks alive without killing them. But Number was young and inexperienced. He got a Moorhen but killed in instantly. He got a second one but killed that one too, so we suspended our hunt. In the evening I tried again with my mist nets and tape recorder, but the Island Cocks were just laughing at me. One of the two dead Moorhens was wearing a ring. It had been ringed by Michael Swales half a year before. It was the first ringing recovery of a Tristan Moorhen in history. A unique event in ornithology.

On Monday I wanted to mark some more albatross nests, but Jeff was impatient and wanted to get back to The Settlement ASAP. He now argued that we had had three days with exceptionally fine weather, and that a fourth, according to Tristan statistics, would be very unlikely. In bad weather we would not be able to take the route along the beach, and because I was such a terrible slow walker, the ugly route across the mountain would take too much time. So we had to depart immediately. But on the way we would take our time, and here and there explore a gulch. It did not happen. Jeff was in a hurry again. Firstly because the wind was going to increase and we could be in trouble at the bottlenecks, and later because it started to rain. With the rain pebbles rolled down the cliff. With more rain, larger stones would fall, making our walk very dangerous. So run, run, run again. In spite of the cooling effect of the rain I arrived home totally soaked in sweat again. On the plus side my two dead Island Cocks could go into the freezer as farm-fresh as could be.

28. Shipwrecks

It was good that we did not stay at Sandy Point any longer as it rained all day. Getting back in the rain along the beach or over the mountain, either way, may have been very difficult, if not impossible. Jeff was right. The village was quiet after the *Agulhas* left, the museum and craft-shop were closed, and the supermarket opening hours had been reduced. Jeff promised me that on the first day with nice weather he would take me to the Base, east of the Settlement, near Pigbite Gulch. We would go Island Cock hunting around Big Green Hill and take a look at the Ponds, three crater lakes on the lower part of the Base at about 700 m above sea level. People would sometimes go to the Ponds on a Sunday picnic.

At Margaret's birthday party sweet sherry and beer were being served.The guests, mostly men, were sitting on chairs arranged along the walls of the living room. Margaret herself was not to be seen. She was in the kitchen, with most of the other women, preparing snacks. When I entered the room all conversation fell silent. When someone asked me how it had been at Sandy Point and if Jeff had been well-behaved (with a grin and a wink in his direction), I answered that it all had been fine and that my only problem was that Jeff was such a terribly slow walker. That generated a good laugh and the ice was broken.

We talked about the weather and fishing. The fine days, during our stay at Sandy Point, had been excellent for fishing but because of the off-loading of the *Agulhas*, it had only been possible on the day after the ship left. We had seen the tiny fishing boats coming around the island from the slopes above Sandy Point. Everybody emphasised that the Island Cocks were really terrible, and if they were not culled, it would soon be the end of the Mollies on Tristan.

During the following days I amused myself with various excursions on the Settlement Plateau, and in the Admin Building I paged through the archives looking in particular for shipwrecks. Jeff was nowhere to be found, regardless of the weather. I wanted to remind him of his promise to take me to the Ponds but he was never at home. As a plumber he had all sorts of jobs to do and simply was never available. I tried Herbert Glass who had made various interesting trips with the Whale People, but he did not want to take me with them because Jeff had been appointed to be my guide and he would not step on his turf. The two ornithologists of the island, Ian Lavarello and Connie Glass, who had both promised to take me to the Molly Plot above The Settlement, were too busy. Connie had to catch up on two years of work and Ian still had a lot of things to do following the visit of the *Agulhas*. Thus, the week went by. Jeff was nowhere to be found and Ian and Connie were too busy all the time.

"Maybe tomorrow," they liked to say. In the meantime I had become friends with Connie's cousin Clive, who had replaced him as a police officer during his stay in England, and now acted as part-time deputy, sometimes even in uniform. Clive's father Herbert owned one of the few dogs which knew how to catch Moorhens without killing them. The dog, Dandy, also listened to Clive, so he would be happy to join me to go Island Cock hunting at The Ponds. When the weather was bad, there was no point in going up, and when the weather was fine he too had other things to do.

"Maybe tomorrow," he said with a smile.

I realised that I had to accept the truth. I was stranded in the Settlement, just like Michael Swales in April. I really believed that this would not happen to me, but there I was, marooned on Tristan da Cunha.

The Education People were having meetings with school personnel all day, The medical staff spent long days in the Camogli Hospital, Roger Roger was happily tapping away, and the Whale People went all over the island with Herbert. I had the days to myself.

From the harbour you can walk along the boulder beach to the west, around Herald Point. There were supposed to be nesting Antarctic Terns and Brown Noddies. I found the Antarctic Terns, but Noddies were not to be seen. The terns were not nesting yet, just resting in small groups, not giving any alarm. Albatrosses wheeled around above the sea and two tiny Diving Petrels whirred past, low over the water. They are called Flying Pinnamins by the islanders. In the distance long strings of Greater Shearwaters were gliding over the swells to their homes on Nightingale.

Southwest of the village, Hottentot Gulch reaches the sea. The gulch, named after the Hottentot regiment of William Glass and Dugald Carmichael, is usually a dry bed, but the size of the boulders and collapsed grass slopes on either side indicate that it must have been a wild torrent not too long ago. On the beach, just beyond the gulch, a white yacht's hull contrasted sharply with the black sand. All valuable parts had been removed, and a large hole was visible on the side. The name was still visible: *Aku aku*, inspired by the books by Thor Heyerdahl. A witness of a new era in the history of shipwrecks. The last disaster with the grand old sailing ships had occurred more than half a century ago. With the advent of steam and fewer ships visiting Tristan, the shipwrecks seemed to have come to an end. But now it was the time for private yachts sailing the world's oceans, and they would inevitably flounder from time to time on Tristan.

The *Aku aku* came to tristan in February 1988 for repairs, but when it left, the keel struck the reef that lies in front of the harbour and the hull was ripped. Solo-sailor Lópes saw no other solution than to ground the leaking yacht. It was dragged high up the beach near Hottentot Gulch, out of reach of the surf. Lópes went to Cape Town, with the intention to come back soon to retrieve his ship, but he never returned.

Tristan has an impressive record of shipwrecks already starting in 1508 with Jorge de Aguiar. In the nineteenth century, since the first settlement, no less than 22 shipwrecks occurred, 5 in the first half of the century, 17 in the second half. The people of Tristan built up a reputation for saving almost everybody every time. Several shipwrecks brought new settlers to the island. Richard Riley from the *Sarah* (1820) stayed and later married one of the Saint Helena brides. The Dane Pederson from the *Nassau* (1825) also married one of them. And when Peter Groen stranded in 1836 with the *Emily*, he married Mary Jacobs, daughter of Sarah Jacobs from Saint Helena.

Ironically, surrounded by all that water, fire at sea has always been a great danger, often leading to abandonment of the ship. This happened to the *Joseph Somes* in 1856, the *Bogota* in in 1869, the *Beacon Light* in 1871, and the *Italia* in 1898. The first three were lost at sea, but the crews managed to reach Tristan in their life boats. The captain of the *Italia* ran his burning ship ashore on Tristan to save his men. The *Italia* brought the settlers Repetto and Lavarello.

Three ships managed to hit Inaccessible by total surprise: The *Blenden Hall* in 1821, The *Shakespeare* in 1883, and the *Helen St Lea* in 1897.

In 1864 the *Lark* broke into pieces on the Tristan cliffs. It was a pirate ship which, during the American civil war, had collected a lot of valuables. Captain Summers and his mate Henderson managed to get a treasure chest on shore and hide it. When it was time to leave they could not retrieve it without giving themselves away so they left it. Years later, after Summers died, Henderson returned to Tristan and removed the loot, which had a value of 35,000 pounds. It made the press and Summers' son found out. He sued Henderson for a share. The court ordered Henderson to pay Summers junior 17,500 pounds.

In May 1878 the American bark *Mabel Clark* got into trouble and became stranded on the reefs of Tristan. Two boys tried to swim ashore with ropes, but they did not make it and drowned. Fortunately the islanders saw the ship. Passengers, including two women, were hanging on to the wreck in great distress. William Green and Cornelius Cotton, managed to swim to the ship with a rope tied around their waists, with the intention to catch those who wanted to jump, and bring them ashore. Some were saved in this manner, but most of them did not dare to jump. Next day the sea was calm, and the islanders came alongside with their boat. Everybody was saved. The heroic actions of the islanders reached the headlines worldwide, and the American authorities decided to send a reward of 25 dollars to each of them. This was never accomplished. Years later, Peter Green received a gold watch, a private gift from the president of the USA.

In 1885 the only shipping accident involving islanders occurred, known as the life-boat disaster. It has always remained a mystery what actually happened. A ship was spotted on the horizon, and the men launched their boat to approach it. From shore, the others witnessed how the ship, which had stopped, started moving again, increasing its distance from the island. The life-boat kept following the ship it until both disappeared. Boat and men were never seen again. Nobody understood why the ship had moved or why the life-boat never returned. The sea was calm and there was no storm. Tristan lost almost the entire male population. On the island, widows and children remained plus one old man and an insane one. Sons who had emigrated returned to the island to help the families get through the hard times that would follow. The ship was the *West Riding*, which reported the incident in an Australian newspaper. They said they saw the boat approaching when suddenly the sail disappeared. When the *West Riding* went to help them, they could not be found. After searching for a while they had to give up. This account does not match the story told by the islanders, and it has been suspected that the captain purposely shook off the boat, to avoid having the burden of an intrusive gang of men. Later, the *West Riding* went down with all hands. God always punishes.

In March 1893 the *Allenshaw*, Captain Thomson, sailed on a course straight towards the island. Shortly before that, the captain had ordered that sails, masts, and oars for the lifeboats should be stored in the hold. He had been hiding in his cabin for days but he kept an eye on the course. When the mate secretly tried to change course by a few degrees to navigate safely around the island, he shouted at him to stay on the original course. What to do? Nobody dared to interfere, afraid to be accused of mutiny. So the ship ran on a reef at full speed, breaking in two. Many reached the shore alive, many drowned, including the captain. His wallet was later found among the debris that was washed up. It contained a letter from his fiancée in which she ended the relationship.

To conclude the era of the great sailing ships and their shipwrecks, there is the mysterious story of the *København*. It was one of the last large sailing ships, with five masts, and a crew of sixty men. In December 1928 it left the port of Buenos Aires for Australia, but never arrived. Search expeditions to various southern islands revealed nothing. Other ships reported a lot of ice on the planned itinerary of the *København*, so perhaps it had struck an iceberg and sunk. A year later, a Tristanian arrived in Liverpool with a different story. In January of the previous year the islanders saw a large ship with five masts. It came from the north, but about seven miles from the Settlement it changed course and drifted to the east, towards an area with dangerous hidden reefs. The sea was too rough to launch their boats so the islanders could do nothing. The ship was clearly in distress, with two broken masts, and it was slanting, the stern deep in the water. It disappeared around the corner behind the cliffs. Later, they found a few pieces of wreckage and the remnants of a life boat, but nothing else. Had the ship gone down or had they managed to get away? Why had the crew not given any distress signals? They had radio equipment. Why had they not lowered their life boats to reach Tristan? Or was everybody on board dead, dying of a mysterious disease? Or had they abandoned the ship in panic and perished? The mystery has never been solved.

The first shipwreck in the new category of private yachts happened in 1953. The aluminium yacht *Coimbra* got into trouble in a heavy gale, five hundred miles west of Tristan. On June 24th, 1953 it ran aground on Big Beach, just below The Settlement. One of the crew of four had been washed overboard and drowned. The other three landed safely on Tristan. One of them stayed for seven months.

On May 27th 1971 The Liberian tanker *Mount Alkis* sprang a leak and it looked like she was going to sink. That would have been an ecological disaster for Tristan. But the ship was kept afloat, pumping, awaiting assistance. The engines failed so it was just drifting around idly. In July the *Statesman* arrived to tow the ship to Cape Town but it didn't make it. Halfway there the pumping could not keep up with the leaking and the ship was left to sink.

In February 1986 the yacht *Chricanto* ran into trouble north of Tristan. On board was a South African family with three children, the youngest of which was seven years old. The name *Chricanto* was composed of their three first names: Christopher, Candice, and Anthony. They had been sailing for five years around the British Isles and the Mediterranean and now they were on their way home. During the night of February 18th they were surprised by a sudden, freak squall which took away the twenty metre high mast. A seven metre stump was left. They managed to construct a make-shift sail. Tristan was 740 miles away. With contrary winds it took them three weeks to reach the island, where they could await transport to Cape Town. A few days earlier they had seen a ship passing by, less than two miles away. It did not give any response to their distress signals of flares, smoke bombs and the like.

On October 17th 1987 Clive Glass saw flares near The Hardies, offshore rocks close to the Potato Patches. A yacht had run aground and was stuck. With the factory launch, solo sailor Siegbert Mrowka was rescued. Siegbert came from Germany and had lost his mast four weeks ago. Yet he manged to manoeuvre the crippled yacht to Tristan. His provisions were down to half a litre of fresh water, one packet of soup and two tins of rice. So he had a narrow escape. After he left for Cape Town the islanders retrieved the wreck on a calm day. It appeared to be nameless but, after scratching away some fresh paint, they found the name *Brandgans* (German for Shelduck). It appeared that the yacht had been stolen in Germany, having been rented for an afternoon in the Waddensea. Interpol tried to apprehend Mrowka in Cape Town, but he had already pinched another yacht and was on his way to Rio de Janeiro.

On January 11th 1993 the yacht *Halcyon* approached the harbour. It had a concrete hull. On board was the retired reverend Kriel from Cape Town, with his son and daughter-in-law. The yacht was damaged in a storm and the Kriels asked for assistance with repairs. Using Judy Green's sewing machine, the sails were patched up in the school hall. Joseph Green and Stanley Swain repaired the rudder. On January 13th, strong winds forced the *Halcyon* to seek shelter in the lee of the island, off Sandy Point, but when the wind died down the Kriels did not come back. The islanders went to look for them in the company launch and found them on the beach. The yacht had disappeared. They said they had anchored and then gone ashore to spend the night in the hut. They had to force the lock to do that. Next morning the yacht was gone, presumably lifted off the anchor and drifted away. The islanders were not amused, especially when it turned out the Kriels had brought a monkey from Brazil without a health certificate. The Island Council decided it was a health risk for the people and the domestic animals and the monkey had to be euthanised and destroyed by the agricultural officer and the doctor. The Kriels were furious and later, back in Cape Town, caused a lot of trouble in the press.

Strangely nough the *Halcyon* was thought to be smashed to pieces on the reefs of Inaccessible. That sounds highly improbable. How could a ship drifting away east of Tristan end up on Inaccessible? But there were the remains of a concrete hull exactly like the *Halcyon*, but a piece with a name was not found. So it could have been another shipwreck and the sailors never found. An unsolved mystery. It was also strange that the monkey stayed on board while the ship was near the Settlement but was taken ashore at Sandy Point, probably indicating that the Kriels had no intention to get back on board. It looked suspiciously like insurance fraud.

The last shipwreck before I arrived at Tristan happened while I was alreday on my way to the island. On September 20th, while I was shopping in Cape Town, the Russian freighter *Polessk* capsized two hundred miles west of Gough. For some reason there was a total failure of all electric systems, making the ship unsteerable. It automatically aligned itself with the waves, which caused it to roll so badly that the cargo shifted and the ship turned upside down in minutes. There was no time to launch life boats. A few inflatable life rafts were thrown into the sea, but these were later found to be empty. Two men managed to put on survival suits, and floated around in the sea for 38 hours before they were picked up by a Malaysian ship. There was no trace of the thirty odd other crew members. One of the two men had been unconscious for a long time and the other kept his head above the water all the time. When they were taken on board the conscious man died on the spot of heart failure, the unconscious one survived. With serious hypothermia, blood circulation on the surface of the body is largely shut down in order to maintain the last bit of body heat in the core. When a victim is pulled up the sudden movement may stimulate blood circulation, whereupon cold blood can flow from the surface into the core and cause death. It is not a coincidence that people often die right at the moment of rescue. An unconsious person does not have that problem. The sole survivor was brought to Tristan where he was revived and pampered for a few hours before he went back on the Malaysian ship, bound for Cape Town. In terms of number of casualties, this was the most dramatic shipwreck so far in Tristan history.

Although not about a real shipwreck, this is the place to tell the story of Janneke Schokker. Janneke was a twenty-year-old student in Groningen, in the north of The Netherlands. In a pub she met a handsome Swedish guy, Sven Lundin, and they were immediately attracted to each other. Sven had built a sailing boat of his own design, the *Bris* (breeze), in his mother's basement in Gothenburg, which limited the length to six metres. He was on his way to the South Pacific but made a stop for provisions and repairs in the port of Delfzijl, not far from Groningen, on the Waddensea coast. In May 1973 Sven and Janneke sailed south, to round Cape Horn to reach the Pacific, but their small boat could not cope with the high waves and strong contrary winds. In bad weather they had to go inside and close the hatch but the cabin was so low that they could not sit upright. In a strong gale the boat made a 360 degree roll. Normally that would rip off all the rigging, but the *Bris* was a sturdy boat and miraculously righted itself undamaged. Janneke was just making coffee (try to imagine!), and thought she had died and arrived in Heaven, hearing angels singing. But it was the kettle flying and hitting the radio button. After days without any progress they had to give up and decided to go east to find the Pacific after crossing the Indian Ocean. But sailing before the wind with such a small boat also turned out to be problematic. In the frequent gales from the west they had to reefe, and when the depressions passed them they would have winds from the east slowing them down. They made another 360 degrees flip over, this time stern over bow, on a huge wave from behind. With slow progress, they ran out of provisions and even worse, drinking water. Their only chance was to reach Tristan da Cunha before they died.

On May 18th 1974 they arrived at Tristan, but due to serious dehydration they were half conscious and delirious and failed to accomplish the last few miles. The islanders had seen them and somehow guessed they were in distress. During the night they left the lights in the village on to guide them but next morning they were still bobbing around aimlessly, but fortunately still in sight. The islanders went to fetch them, and towed them into the harbour. The *Bris* was hoisted onto the quay and Sven and Janneke lived there, waiting for winter to pass. Sven stayed four months until on a nice day he was able to leave, promising he would only head north. Janneke had left Tristan earlier, finding passage to the Cape. She had had enough of sailing.

The *Bris* is in a nautical museum in New York. Sven changed his name to Sven Yrvind, and built the *Bris II*, even smaller, only 5.90 metres long. He managed to sail solo around Cape Horn in winter time, earning him the 1980 Royal Cruising Club Medal, for rounding Cape Horn in the smallest boat ever.

29. Fishing Day

Every morning at six, two men get up to watch the sea and the sky. Together they walk to the Lobster Factory above the harbour. They carefully study the sea, the waves, the wind, the clouds. Then they look at each other and usually shake their heads in silence. No fishing day again. Words are not needed. There is too much swell from the north, or high clouds indicate a depression coming. The sea may look like a mirror, but that can change within a few hours. Last week a fishing day had to be aborted by noon, and at two a full scale gale was blowing. The two men looking at the conditions have to be unanimous in their judgement. If they disagree there will not be a fishing day. On average there is a fishing day every three or four days. In summer it is a bit more often, in winter less frequently. There is no fishing on Sundays, the Day of The Lord. All in all, there are usually about one hundred fishing days in a year. Of course this only applies to the small dinghies operating around the main island. The factory vessels *Hekla* and *Tristania II* fish every day, regardless of the weather, but they operate only around the outer islands.

The sea around Tristan is reserved for the islanders. They fish in fixed couples of two men. There are twenty dinghies, so there is place for forty fishermen (remember this is the 1993 situation I am describing). On non-fishing days, of course, every fisherman has other jobs to do but fishing always has priority. In addition, there are young men learning the trade, joining as a third man on board, or replacing the second one, like Graham Rogers, Jeff's son, who had just finished school.

When the six o'clock judgement is positive, the men bang on a free hanging empty gas bottle, a giant ding-dong resounding all over the village (also used to announce distribution of mail after a ship has arrived, but this never happens at six in the morning.) From everywhere you see men in blue overalls appearing, walking towards the

Lobster Factory. There they collect the frozen bait they are going to use, often octopus or fish remains. From the factory they drag the bait to the harbour in sledge-like plastic boxes, loudly scratching over the concrete. There is hardly any talking. Everybody knows exactly what to do, no words are needed.

It was a fishing day and, as I could not find anyone available to take me on the mountain, I asked if I could join. I could come as a passenger on the company launch, but had to sign a declaration that I did this entirely at my own risk. Fine.

In the harbour, the boats were prepared. Bait was placed in the lobster pots and these were piled up in the boats, each attached to a rope with a coloured buoy on the other end. Piling up had to be done very carefully to avoid the ropes getting tangled. It was a colourful spectacle, the yellow boats, the blue overalls, the buoys in different colours, and the deep blue sea in the background. And still very few words were spoken. All boats were resting on small lorries. One by one they rode to the harbour, where a large crane lifted them up and deposited them in the water, usually with the engine already running. In the old days, the boats were made of timber from shipwrecks, covered with canvas, and rowed with man power. Now they are made of plastic, fibreglass and wood, with a noisy diesel engine. One by one, they left the harbour, one to the left, one to the right. On the cliff above the harbour, the secretary of the factory was keeping a list, jotting down the time and the direction of every couple leaving the harbour. This time, eighteen of the twenty boats sailed. Usually four of them have a walkie talkie on board to keep contact with the company launch, which has a stronger radio to communicate with the factory. In this way, everybody can be warned in time when the weather deteriorates and the fishing has to be aborted. But most fishermen do not need a warning signal. They can judge for themselves when it is time to quit.

The company launch was the last to leave the harbour. On board were skipper Barton Green, and his mate André Repetto. We turned left, towards the Potato Patches. From shore the sea looked like a motionless mirror, but being on it in a small boat, there appeared to be a lot of movement, with a long, steady swell, causing us to roll around heavily. Every now and then a splash of spray would go over us and I had to take great care to keep my cameras dry, seeking shelter in the small, half open cabin, much to the amusement of Barton and André. André spent much time sitting on the bow deck, leaning against the window, blocking Barton's view. He would get a shower at times, but would soon be dry again, it being a wonderful warm sunny day.

The air was brilliantly clear and for the first time Inaccessible showed structure, instead of being just a dark lump of rock, with clearly visible scars of landslides. The sea was just as clear as the sky, blue at a distance, but looking down into the deep, it was pure black. The world around Tristan is a very clean one.

Tristan's lobster fishery is strictly regulated. Every fisherman has to use a fixed wooden imitation of verniers callipers, indicating the minimum size of the carapax. Any crayfish falling within this size has to go back into the sea immediately. Females carrying eggs are also thrown back. A second check takes place when the animals are dumped on the conveyor belt in the factory. The Agricultural Officer measures them all over again. Usually hardly any undersize lobsters are found, but they are ceremoniously brought to the harbour to be released. Having been out of the water for hours already, their chances of survival are minimal, but is has merely to be seen as a symbolic gesture. If undersize animals are allowed to be processed the fishermen at sea will become less accurate.

In 1993 the concession was in the hands of Tristan Investments Ltd (PTY), Cape Town. Later it passed on to Ovenstone Agencies (PTY) Ltd, also based in Cape Town, with even more strict regulations to prevent overfishing, such as an annually determined quota. The number of boats fishing around the island has been reduced to nine. To keep the fishery sustainable is absolutely vital for the future survival of the people of Tristan da Cunha. In 2011 the Tristan fishery obtained the Marine Stewardship Council (MSC) award.

The lobsters caught by the Tristanian fishermen are processed in the factory on shore, but the catches by the ships are processed on board. Most tails are frozen, as in the factory, but a small number of animals are cooked whole, especially for ornamental dinners. This involves more work and the product is more voluminous per meat weight, but a much higher price compensates for that, All lobsters are for export. They are not for sale on Tristan. So anyone who wants to taste them needs to be invited by an islander, which of course is never a problem with all those open doors.

As soon as the fishermen have found their fishing spot the traps go overboard, thirty, forty metres deep, in the kelp zone. A brightly coloured buoy marks their place. The boats are moored by just dragging some strands of kelp into the boat and winding them around anything. The same technique is used by the Sea Otters of the Californian coast. The traps stay down for three hours to allow the lobsters to walk into them and then they are hauled up. On most days this happens twice. During the wait the men try to catch some fish, using a simple line with a hook, for their evening meal, or to be used as bait on the next fishing day.

After visiting a few fishing dinghies we reached Barton's favourite fishing spot. Although we were five kilometres away from the coast, he knew exactly where we were by using land-marks. The art is to have two land-marks in the same line of vision, and do the same trick with two others, a simple way of triangulation. Barton knew that this was the best spot for Bluefish, an endemic Tristan species, and the tastiest of the whole Atlantic. The distribution of

Bluefish is rather irregular, depending on the properties of the seafloor.

We were southwest of Burntwood, with views of the Settlement Plateau and The Caves. The little white huts of the Potato Patches contrasted sharply with the dark background of the mountain slope. Above the Base, the eternal clouds obscured the peak all day, but both Nightingale and Inaccessible were sharply visible under a cloudless, clear blue sky. Mollies were gliding around us all the time, and often small groups of Greater Shearwaters passed to and from Nightingale. We saw several Diving Petrels, with their whirring flight. Their wings are too small to glide and they have to flap at high speed, like the auks of the northern hemisphere. They look very much like the Little Auks of the arctic Atlantic. The first Norwegian whalers arriving in the south really thought they were the same. A beautiful example of what is called convergent evolution. Both the Little Auks and Diving Petrels use their greatly reduced wings to 'fly' under water when diving, just like penguins do. No wonder the islanders call them Flying Pinnamins.

We used no rods for fishing, just a rather thick nylon line with a large hook, baited with a chunk of octopus. The line was half a millimetre thick, the hook was ten centimetres long, and the bait weighed at least a hundred grams, so we were not fishing for sticklebacks or anchovies. Barton let his hook go to a depth of twenty metres, and hauled it up again immediately. He caught a Bluefish, half a metre long, with not only the bait in its mouth, but also its air-filled swimming bladder, which had come out because of the difference in pressure of about two atmospheres. André also caught his first fish immediately. Only my line remained empty for quite a while. I could not understand what I was doing wrong, but both Barton and André were highly amused, catching one fish after the other all the time. Suddenly, they too were out of luck. The current had carried us away from Barton's spot. He started the engine, and went back to the right place. This time I caught a big Bluefish myself, and soon a second one, giving me blisters on my fingers hauling up the line. I also caught a huge Steenbrass. Also delicious, so I was told, but you have to eat them fresh, otherwise they soon turn oily. Bluefish meat is lean, but Steenbrass is rather fat, making it very good for smoking and cooking on the barbecue.

Again the hooks came up empty. Barton did not understand why because we were certainly still at the right spot. And then he hauled up a large shark. André came to assist, hammering the beast on its sensitive nose with a large piece of wood. They did not bring the fish on board, but with a sharp knife they slit its belly open over the full length, and let the animal go, its entrails trailing behind. Feast for the others down there. When I asked why they treated it so cruelly, their response was simply:

"Man and shark have been created as enemies."

Not an opinion, just a given. I have seen the same reaction in the Mediterranean. With an excursion of biology students, we were on a small fishing vessel to admire the great variety of Mediterranean sea life. We saw sea turtles, interesting seabirds, and all sorts of colourful underwater life. Once, the net brough two small sharks to the surface, not more than one metre long, with a beautiful blue sandpaper skin. The fishermen got all excited. With a huge pair of scissors they removed the fins and the tail, and they cut open their bellies and threw them back into the sea, their entrails dragging behind. The clumsy way in which the finless creatures tried to get away caused great amusement. When asked why, they came up with the same answer: men and sharks had been created to be arch enemies.

Now Barton knew why we had stopped catching anything. When sharks arrive the Bluefish go away. My line came up without a hook, the line neatly cut. Also a shark. We gave up and moved towards Stony Hill.

Near Stony Hill a huge tail was sticking out of the water. A Southern Right Whale was playing, perhaps the same individual I had seen yesterday near the Settlement. He was hanging upside down in the water, slowly waving with his tail in the air. This is a well known behaviour of the Right Whales which is not fully understood. Often they do it in shallow water, so it has been suggested that they scratch their itching head on the rocks, but this one was performing in deep water. I wanted to have a closer look but Barton was a bit wary. Whales are friendly animals but what if he decided to scratch his head against our boat? It would not be the first time a Right Whale upturned a small boat. It was said to have happened right in front of the Settlement. Everybody was able to get ashore safely, but Michael Repetto badly grazed his thighs after he landed straddle-legged on the Whale, riding it astride for a while.

Barton approached to about a hundred metres. The Whale stopped tail-wagging and came to the surface full length. The head was fully covered in white patches of what I thought to be barnacles, which grow on many Whale species. Much later I learned that the Right Whales do not carry barnacles, but large colonies of small crustaceans instead, which are called Whale Lice. Like humans, they host three species: one on the surface of the callouses on their heads, another species in the clefts in those callouses, and the third living in the genital slit. Southern Right Whale and Northern Right Whale each have their own three species. Our whale loudly blew a steamy cloud into the air, and slowly sank, like a submarine. Later, we saw the tail-wagging behaviour again, far away.

On the east side of the island we had a look at some of the fishing boats. On the cloud-covered mountain it must have rained, because suddenly white waterfalls appeared all along the cliff's edge. The area between Stony Hill and Sandy Point looked very wild and inhospitable. This is where most of the Mollies nest, and where people hardly ever venture.

Sandy Point, Big Point, the new volcano, and then the harbour, two pm. The launch was going for a second round, but first dropped me off, giving me time to clean and process my fish and be back in the harbour in time to see the boats coming back. I filleted my fish and cut them up in one person's portions for the freezer. At six pm I was back at the harbour, where the boats came in one by one. Almost all the women were involved in processing the tails. The legs (there are no claws) are considered offal but contain just a little tasty meat. Noe and Pat were scavenging around the factory, assembling as much as they could. For us, a fishing day meant nibbling legs for dinner. Noe was especially proficient. Where others managed to suck out the meat of ten thin legs he could process a few dozen.

The lobsters are assembled at sea in a large net which can be hoisted up immediately after arrival in the harbour, where it is weighed. Fishermen get paid a fixed amount for a fishing day, plus a bonus per pound. Three hundred pounds is considered a good catch. Weighing was done by Lars Repetto, father of André, and book-keeper of the factory. He was born on the day the Norwegians arrived in 1938 and named after Lars Christensen who had organised the expedition.

In the factory the catch is landed on a conveyor belt, where six men with sharp knives cut off the tails. The breast parts are thrown into wheelbarrows or land on the floor where they crawl around aimlessly. Wheelbarrows are emptied periodically and the offal is processed to pulp, which is returned to the sea via a pipeline. Part of it is left to rot in barrels, to become fertilizer for the Potato Patches.

The tails land in a steel gutter with flowing water, where they are picked up one by one by a group of women who remove the black vein (actually the gut), which is said to be poisonous, with a small knife. Then they end up in a rotating steel drum, the washing machine, and then in a large steel basin filled with water. Two men transfer them into smaller tanks and bring those to the heart of the factory: the packing area. A whole army of women is busy sorting, weighing, and packing the tails ready to be frozen. They are all dressed in white aprons with blue caps on their heads and impeccable white boots on their feet. The area is strictly off-limits for everbody not working there. This is a matter of hygiene, but also to prevent tails from secretly disappearing. One of the women is not taking part in weighing and packing but standing aside, supervising the whole process. Tristan style, this is not a permanent job. Next time another woman will stand there. As the freezing plant will have to run full power all night, the whole village will have electricity all night too so it is party time everywhere.

We failed to secretly secure a tail or two so we had to be content with the legs.

30. The wedding

Saturday. The wedding of Martin and Iris (pronounced 'Hoiris'). Having the same surname, Green, everybody will ask how closely they are allied. In the straight line of fathers of fathers, we have to go back to Peter Green to find a connection, six generations back for the one, five for the other. Switching to female lines, we get there earlier. Their nearest connection is one shared great-grandfather. So they are not all that closely allied. My father's grandfather and my mother's great-grandfather were brothers, and their wives were sisters. I am my mother's cousin, and my father's uncle, and my own nephew or great-uncle. So I am a worse case of inbreeding than the children Martin and Iris will produce. This example illustrates again how shamelessly we, interfering outsiders, can dig around in the Tristan family trees.

The marriage was announced on the notice board of the Admin Building, Martin being categorised as 'bachelor', Iris as 'spinster'. Martin had an agricultural job and was a fisherman, Iris was secretary at the financial department. On the form they had both signed for being of age, not needing anyone's consent, being members of the Church Parish of Tristan da Cunha, and not knowing of any legal objections to their marriage.

Much has been written about Tristan marriage customs. In the old days a prospective suitor would visit the home of the girl of his choice and sit silently with her parents, not even looking at her. When he had done that three times he was accepted as a partner for the girl. He would give her a pair of self-made moccasins, and she would give him a pair of socks in return with the famous barcode indicating her love. After exchanging moccasins and socks the couple was considered to be engaged.

Tristan houses offer little privacy for young lovers and making love outside in the bushes is often not much fun because of the weather. Nonetheless, many couples managed to conceive a child before the wedding date. Waiting for a wedding date often depended on finding a home. Sometimes they would build an extension to the their parents' house. Building always depended on the help of friends and that could take a lot of time, because friends would rather help tomorrow than today.

Times have changed. Boys and girls just meet and dance in the disco, like anywhere else in the world, and boys no longer make moccasins for the girls. Girls may still give socks, but the barcode has lost its significance, the stripes (usually blue) are purely ornamental. Getting pregnant before the wedding still happens frequently. Martin and Iris had a two year old little girl already.

It was a gorgeous day. No wind and a blue sky. Clearly a fishing day, so many men would not be able to attend the wedding in St. Mary's Church. There was no reverend on the island so the ceremony was led by lay brother Eddy Rogers. Guests were arriving, finding a place in the church. All the men wore black suits and glossy shoes, women wore dresses and high heels, the elderly ladies had colourful kerchiefs. A fifteen year old girl played happy tunes on the harmonium. The bridegroom and his family stood outside, waiting for the bride, who was late. She arrived in the Admin's white landrover with her parents and the bridesmaids. Bride in classic white, bridesmaids in identical orange flowery dresses, with matching hair ribbons and flower baskets. The whole ceremony was utterly ordinary, like in any European village, no special customs or island folklore. Psalms were sung, there was a sermon, the couple repeated "I do", and exchanged rings and there was more singing. Then the religious part was done and the Admin added the formal civil part, the witnesses had to sign, and that was that. The newly-wed couple left the church looking as though they were glad this was all over.

Outside it was so nice and warm that nobody was wearing a coat. In front of the church, pictures were made of the bridal couple, with and without parents, with and without witnesses. Then they stepped into the landrover, which had been decorated with balloons and flowers and the words JUST MARRIED on the rear window. The landrover moved away and disappeared in the village. That was a strange moment. Suddenly the Tristan wedding became special after all. Where to go on Tristan after the wedding? Honeymoon at the Potato Patches? It appeared they went for a ride through the village, to have themselves congratulated by the older peolpe who had not been able to come to the church. A nice custom. Meanwhile the guests gathered in and around the brand new house of Martin and Iris, where family and friends had assembled an enormous quantity of pastries in all sorts of sizes and colours. The bridal couple arrived too, glasses were filled and we all toasted and sipped our drinks. With a beer in the sunshine and not even wearing a sweater, this was far from being a subantarctic experience.

In the evening there was a dance in the Prince Philip Hall. At first all the boys and men sat stiffly along one wall, the girls and women along the opposite wall, but during the course of the evening the atmosphere lightened, and everybody was soon dancing with gusto. The older people danced the classic way, the youngsters disco style, like anywhere else in the world. No more folklore. There was live music by the Triss'n Rocklosters for a while, but most of the time there was deafening pop coming from ghetto blasters.

After a while I had enough of the noise, and I fled to find fresh air and silence. High up in the air petrels were uttering their weird cries. I walked towards the new volcano, where the birds were supposed to have settled, but the sounds gave no further indications. Walking back into the village I saw a strange light, caused by a torch lying on the ground, next to a passed-out person, feet up in a ditch.

"Oh, don't worry, she will be all right, just had a few too many," a voice said, behind me. It was the local police, beer in one hand, emerging from Joyce Hagan's house.

"You want one too? Just go inside and grab one."

It appeared I just fell into the 'tea break'. Inside the house it was crowded, and there was a lot of laughter. I stayed there the rest of the evening. Only the youngsters went back to the Prince Philip Hall for more dancing. Late at night I walked past the Lobster factory. I still heard petrel cries in the dark sky. It had been a fishing day, so the women in the factory had to work till late in the night, cleaning the tails and preparing them for being frozen. The village would have electricity all night. I recognised the faces of several people, including the bride's mother. Life goes on in Tristan.

31. The Ponds

The day after the wedding was another beautiful day, with a blue sky and bright sunshine, but the wind had freshened and whitecaps were running into the harbour. No fishing day. I had enough of hanging around in the village and wanted to go on the mountain. But it was obvious I would find nobody to act as guide again. It was a Sunday, people were still tired after yesterday's party, so maybe tomorrow. I only saw one solution: civil disobedience.

Accoring to the map the trail across the mountain to Sandy Point started at Pigbite Gulch, not far east of the new volcano. In most places the slope was incredibly steep, ending in a sharp edge against the blue sky. Without a trail it would be impossible to get up there. From the beach no beginning of a path was visible. I continued far enough to the east to be sure to have the trail behind me. Then I climbed the grassy slope diagonally back to the west, where I would meet the cliff. Thus it was inevitable I would cross the path. After I found it, the rest of the climb was simple and easy. Apparently the trail was used often enough to keep it maintained. There were tracks of humans and sheep, but no fresh ones. In a few places it looked like the path split and I had to take care to memorise the lay of the land. In a few steep places ropes were attached to the rocks, to assist climbing, but they looked old and withered, so I preferred to trust my own hands and feet.

On a steep fern slope I suddenly found myself face to face with a Moorhen. I tried to lure it in my mistnest with the sound from my tape recorder but it did not respond. Before I reached the rim I met four Moorhens. How nice to be here without a dog! The last metre I hoisted myself onto the edge, and there I was, on the plain of the lower Base, rather flat, slightly undulating, with a vegetation of tree ferns that were much smaller than those at Sandy Point. In front of me was Big Green Hill, a neat little secundary volcano with a flock of grazing sheep. Behind me were fantastic views of the ocean, the Settlement, and the new volcano. From where I was standing the path down was totally invisible, so I really had to make sure to be able to find it on my way back. I memorised strangely shaped trees and other landscape marks and took a compass bearing on the peak of Big Green Hill. Thus, I would be able to find the path even if I was surprised by fog. I walked back and forth a few times, between Big Green Hill and the edge of the cliff, looking at it from different directions. There was not one single clear path here because the sheep had created a whole network. Now I could move around on the Base freely with Big Green Hill as a fantastic land mark.

I found my way towards the Ponds, sometimes using sheep's trails, sometimes just fighting my way through the ferns. I had to cross a few narrow, steep gulches, where I had to find a spot where I could lower myself between the ferns and climb out on the other side, using the ferns again. I met another Moorhen. This one was scratching the earth like a chicken, clearly looking for food. All I could find was woodlice, so that was probably what he was after. He was certainly not looking for albatross eggs.

The ponds were very scenic, three in a row, connected by beautiful little waterfalls. The lowest of the three had no exit so the water had to disappear into the porous mountain. Ink-black water, surrounded by incredibly steep green slopes. Wind gusts produced the strangest patterns of ripples on the water. Between the two upper lakes I could walk along a dam, close to the shore of the one, high above the next, easily stepping across the connecting waterfall. The higher parts of the mountain and the peak were invisible. Above 1200 m whirling masses of black clouds were blown away, while new ones were forming. The weather up there was not good, but at the Ponds the sun was shining. On my way to the Ponds, I repeatedly looked back and took my compass bearing to Big Green Hill, to make sure I could find it, whatever the weather had in store.

Between Big Green Hill and The Ponds I met sixteen calling Moorhens, three of which showed themselves. This area has only been colonised by them in the last ten years. When Michael Swales was here in 1983, with the Denstone expedition, the birds had just reached the gulch east of the Ponds. They have now extended their area up to Hottentoch Gulch, above the Settlement. Above the upper of the three lakes I found another sheep's trail, which I could easily follow back to Big Green Hill, in a wide arc, slightly higher on the slope. On the way I counted twentyone Moorhens.

32. The Caves

Monday after the wedding. Already at Sandy Point Jeff and I made the appointment to go to The Caves today. He would fetch me in the morning, and then we would walk past the Potato Patches to the end of the Settlement Plateau, where we would climb onto the Base, to reach the southwest point of the island. Like at Sandy Point, there is an area of grassland of a few dozen hectares where semi-wild cattle graze. There are a few huts where people can spend the night. The great attraction of The Caves is a small herd of Subantarctic Fur Seals, which is now slightly increasing with protection. Some islanders complain that the seals are starting to lie around in the grass, ruining it for the grazing animals. Farm people are the same all over the world.

Before Jeff arrived I went to his house to check if all was still on schedule. We had our appointment, but during the past week lots of other appointments came to nothing, so I wanted to make sure. I had been to his house the previous evening, but everybody had gone to bed by 9 pm. So now I wanted to be there early, before he could escape to do other things. I was too late. Jeff had already left for a plumber's job, leaving a message for me that we would go the next day. I did not like that, because it was another fine day (a fishing day again), and with postponing things on Tristan you never know what the weather will do. So I pressed on. Jeff's job would be done at 10 am, so I managed to pursuade him to go after all. He would come and fetch me at 10:30. At 10:45 he came, running, and said he had to do yet another job. He told me to walk to the end of the Settlement Plateau and wait there for him. He would catch up on his motor bike. Fine, at least I could now walk at my own speed. I got a lift from someone driving to the Potato Patches. People were digging potatoes, some even with bare chests. On sunny days Tristan is much warmer than all the books would have you believe.

At the end of the Settlement Plateau a slope of bare red gravel goes up, leading to a steep green slope with grass and ferns where in the past people have removed all the trees for firewood. There has probably been a forest fire too because the place is called Burntwood. It is said to be the easiest way to get up to The Base. It is only a 500 m climb and is not very steep. People used this place to drag collected wood from the Base down to the Settlement Plateau. Deep ruts in the grass are still visible as silent witnesses. If from this point on you wish to go further south, the only way is up, via the Base, as there is no passable beach at the foot of the cliffs.

From a vantage point a little higher up I watched Jeff approaching with his motor bike. He parked and immediately started running up. We were late, and we were in a hurry. In thirty minutes we reached The Base, drenched in sweat again. No time to rest, we had to push on. Black clouds were whirling around the mountain higher up, and if it started raining, the gulches could turn into wild torrents and we could easily get trapped between two of them. This is the wettest part of the island, where the strong wind keeps the ferns low, and trees only survive in sheltered gulches. Peat moss proliferates, on the other hand, and large parts of the trail were just swamp. Jeff replaced his trainers with the cut-off feet of a diver's suit to keep his feet dry. My mountain boots were just high enough to keep the water out when I sank into the mud ankle deep. This area is called Soggy Plain.

Above the sea a thin cloud layer drifted towards us, changing into dark, thick clouds when it hit the mountain. Under the light veil the sea coloured pale blue, silvery in the distance. Nightingale and Inaccessible were clearly visible. From where we were we saw the sea behind Nightingale, over its hills, so it was hard to imagine that the island was about forty kilometres away.

We had to cross three large, deep gulches, aptly named First, Second, and Third Gulch. Each time we had to descend roughly a hundred metres and climb up again on the other side. From a burrow, concealed behind a vertical moss wall, the purring call of a Broad-billed Prion emanated, like the call of a Turtle Dove. In sheltered places we also met a few Starchies, the Tristan Thrush. It is strange that these are so scarce on the island. They are very numerous on Nightingale and Inaccessible, and they used to be very common on Tristan too in the past. Every now and then we saw a Molly, sitting on its nest, but on this side of the island there are not many left. Their chicks tasted too good. We neither saw nor heard any Moorhens. They had not yet reached this part of the island.

Via Gypsy Gulch we descended to the small grass plateau near The Caves. Sometimes we slid down over large bare boulders, sometimes we had to climb out again, to avoid vertical dry waterfalls, and cling to steep fern walls. Once we followed a narrow ridge, where a row of Mollies was sitting, their heads into the wind. This must be like the birds Carmichael kicked into the ravine in 1817. On steep grass slopes Number got very excited again, sniffing at Haglet burrows.

The first cow we encountered was dead, a half born calf hanging out at the back, also dead. The second and third cows we saw were alive and grazing, but the fourth was dead, having also died in childbirth. In the distance we saw another dead cow but eventually there turned out to be many more alive and well. Jeff could recognise his own cows, and to his great relief they were all healthy. At Cave Point, there is a small group of huts, like at the Potato Patches, where you can spend the night. Jeff does not own one, but he got the key of a friend's cabin. Inside, it was incredibly damp and musty smelling, fungi growing everywhere, all the household utensils covered in mould. The ideal place to become rheumatic. But there was a fireplace and a pile of firewood, so with a nice fire to chase the humidity, and a fine shot of whisky, life at The Caves was not so bad after all. In the embers I cooked my steenbrass fillets, wrapped in aluminium foil, while Jeff made the usual Tristan Chips.

Next morning Jeff wanted to make a quick dash to Stony Hill to count cows, so he did not want my slow company. I stayed at The Caves and had a good time with the seals, which were playing in the surf and climbing the slippery, seaweed-clad rocks. The seals were not really occupying the grass, but were just lying around on the transition from rock to sparse vegetation. A big bull was lying in a cartoonlike fashion draped over a big boulder, with a fantastic happy smile on his face, and closed eyes. Enjoying life. The youngsters were all engaged in serious battles. They all wanted to become like the big one too, treated with respect by the others, and owning all the girls.

I don't remember why we had to make the return trip running again, but there was a reason, of course. When Jeff zoomed away on his motor bike, I enjoyed a leisurely walk along the cliffs of the Settlement Plateau. I finally found the tropical Noddies, nesting side by side with the Antarctic Terns.

33. Hoil'n Cock

Hoil'n Cock is Tristanian for Island Cock. I heard the word many times during my visits to people's homes. Interestingly enough, everybody knew about the Moorhens stealing albatross eggs, but of all the knowledgeable people I have talked to, nobody ever saw it happening. They all pointed to another eyewitness, and that one to yet another one, going around in a circle. Broken eggs, obviously picked at, have been found every now and then, but that also happened before the Moorhens arrived. Then, the Skuas would simply be blamed. But now, people say, sometimes an egg is found with only a small hole in the blunt end. That cannot be explained by anything else but a Moorhen bill. Can it? Moorhens have been seen curiously circling albatross nests, and one was even observed standing on top of a deserted nest. Lots of circumstansial evidence, but no real proof.

These observations can be explained differently. Gough Moorhens have been studied on Gough. They are scavengers, inspecting petrel burrows, looking for abandoned eggs or dead chicks. Stomach contents showed petrel bones and egg shell fragments. They also frequent albatross nests to look for spilled food, like squid or fish. And a broken egg, or the remains of an egg predated by a Skua, they will certainly feed on. But on Tristan a deaf ear is turned to such stories. Even if nobody ever saw it happen they all know how it is done. It is a piece of island culture that, as an outsider, you have to stay away from. This phenomenon is not unique to Tristan. Friesland, the northernmost province in The Netherlands, has a long tradition in collecting Lapwing eggs, which is controversial nowadays. But don't you dare, as a non-Frisian Dutchman, to say anything about it.

This outsider feeling, the outsider label, was given to me on one of the very first days on Tristan when we were all invited for a reception at the house of Chief Islander

Lewis Glass. His wife prepared delicious lobster bites and home-made Tristan crisps. Inevitably, conversation included Moorhens. Lewis confirmed once more how devious and clever these birds are, operating in pairs, the one distracting the albatross, the other one taking the egg. Had he ever seen something like that? No, but he referred me to other eyewitnesses.

Then we talked about the Mollies, which were eaten in the past but were now fully protected. I told the story of Dugald Carmichael, kicking albatrosses off the ridge left and right, and throwing one into the abyss, where it could not regain flight and dropped dead. Lewis did not believe that story.

"I have seen those things," he said, with a devious smile, clearly meaning that he himself had thrown albatrosses from a cliff. None of those ever fell down, they all just flew away. My story was only good for the rubbish bin. The look on his face: you poor outsider, you don't know anything about these things. My companions fell away from me, choosing Lewis' side.

"Here's someone who knows what he is talking about," Pat laughed. I could have tried to explain that I was not airing my own opinion, just quoting an early eighteenth' century source, but I let it go, it was useless. We talked pleasantly about other topics and Lewis has been very helpful in introducing me to other Moorhen connoisseurs.

With Joyce Hagan I had a different Hoil'n Cock project. We had agreed that she would knit me a sweater, not with the obligatory Tristan silhouette plus albatross, but a special design of my own. She gave me a piece of squared paper on which to draw my design. What else could it be but an Island Cock? After she finished mine she must have produced more. Years later I saw her Island Cock sweaters for sale in the craftshop. I should have asked for royalties.

On murky days I started to deliver the *Tristan da Cunha Association Newsletters* to Tristanian members. I went first to Herbert Glass, the most knowledgeable man on

Island Cocks, I was told. He was not there as he was on the mountain with the Whale People again. His wife offered me tea. He was not only the best Moorhen catcher, with his softmouthed dog Dandy, but he also came nearest to being a true witness. He once found an egg with just a small round hole, right after he had seen a Moorhen near an albatross nest. He took the egg to show it to the Admin. Where the Pigbite trail goes over the edge of the Base he hid the egg under a fern because he had a sheep to shear. After that he forgot about it.

Harold Green offered me tea. He had been Chief Islander a few times, knew a lot about nature and was a guide on Gough with the British expedition, and on Inaccessible with the Denstone boys. He had two definite opinions about Island Cocks. One: it is absolutely true that the birds steal eggs. Two: nobody has yet seen it actually happen, not even he himself. Being over sixty, he just remembered the wooden shoes arriving in 1937. As a little boy he had walked in them but he did not really like that.

Repetto faces are among the most typical Tristan faces, recognisable in all old pictures. A long, angular face with pronounced ears and a large, hooked nose, and a very dark complexion. These characteristics, however, did not come from ancestral father Repetto himself, but from his wife, Frances, granddaughter of pale, round-faced Peter Green, so the 'typical' Repetto features come from Saint Helena. The brothers Ernie and Michael clearly had this Frances-look, just like their sister Millie, who travelled with me on the *Agulhas*. Ernie's grown up children have left the island. He was an exception having no opinion about Island Cocks. But he had a strong opinion about the wooden shoes. They were terrible, useless. On a wet day in the garden you could use them perhaps, but walking in them hurt badly and you could easily sprain an ankle. On rocky soil the soft wood would quickly wear, so they were all gone within two years. But, so he whispered in my ear:

"They made lovely firewood."

Sidney and Alice Glass offered me tea. Their small

living room was, like in other Tristan houses, lined with masonite. Glossy paint. They were both very old, and Sidney had trouble breathing and talking. They had not been on the mountain for a very long time, so they left the Moorhen problem for the younger generation. But Brander's wooden shoes they remembered very well. Great fun! There was a huge pile in assorted sizes and you could just take your pick. They liked walking in them, with nice dry, warm feet. Useful around the house and in the Potato Patches but not good for real walks. And they wear out quickly, so in one or two years they were all gone (or perhaps burned by the Repettos). Alice showed me pictures she had received from another Dutchman. In April, this year, retired whaler Albert Veldkamp from Vlissingen was here with his wife, on the RMS *Saint Helena* (which rescued Michael Swales after seven weeks marooned on the island). Alice made them lunch but then there came a hoot from the ship. Because of increasing wind everybody had to be back on board as soon as possible. Alice quickly packed the lunch in a bag. In the picture she showed me Veldkamp and his wife were enjoying her goodies in their cabin on board.

In the house of Nelson and Winnie Green (glossy blue masonite walls and ceiling) I also found traces of Dutch visitors. There was a signed sketch of the yacht *Hadewych*, in which Eerde Beulakker and his wife had visited Tristan a little while ago. Here our paths crossed for the second time. Two years earlier he had sailed from Elephant Island, in the Antarctic, to South Georgia, in Shackleton's wake. I was on Elephant Island just before him, but here on Tristan he beat me to it.

The Nelson and Winnie names have nothing to do with the Mandela's. They were both in their seventies and had been married almost fifty years, long before anybody had ever heard of the other Nelson and Winnie. Nelson was already over sixty when he joined the Denstone boys on Inaccessible and he was Mike Frazer's best friend. Knowing nature, he was convinced that Moorhens take eggs, but his judgement was rather mild and he did not believe it was a real problem. When I brought up the wooden shoes they both laughed. Yes, they were young, and they had a lot of fun with them. Useless, but fun.

Nelson's garden looked neat with beautiful geraniums in flower. His great pride was a non-flowering plant with large, glossy green leaves: *Peperomia tristanensis*, the Tristan Pepper Tree, one of the rarest plants in the world, only to be found in the Chilean Juan Fernandez Islands and on Inaccessible Island, where they only grow in one small valley on the western slope, no more than a handful of plants.

Some Association members offered me tea, others a beer. I had had five beers already when I came to the last of the day, Neil Swain, the shepherd. He had a friend visiting. They had assembled a pile of empty beer cans around their chairs and I soon started building mine too. When the conversation came to Island Cocks, he turned red and became very angry. It was all Michael's fault. First introducing the monsters, and then coming up with intellectual chit-chat that the birds were not so harmful. Come on! Everybody knows they take the eggs! Island Cocks do not belong on Tristan.

When I told him that in the previous century there had been Moorhens on Tristan, he exploded. First Michael, and now me with my useless diplomas, talking nonsense about Island Cocks.

"There's no Hoil'n Cocks on the hoil'n before Mister Swales put them there!" he yelled.

"My father is ninety years old, and he can tell you there never were Hoil'n Cocks!"

I tried to explain that I was not talking about his father's time, but a century earlier, but now I had done it. Mister know-it-all, stubbornly insisting on knowing better. I still had to learn a lot about diplomacy, my fault.

"There's no Hoil'n Cocks on the hoil'n before Mister Swales put them there!" he yelled again. "Shouldn't 'ave dunnit!"

240

I agreed on that one. He certainly should not have done it. Messing around with species on remote islands only spells trouble.

We talked a bit on various other topics and Neil calmed down. I declined the next beer and left. From the kitchen his wife gave me a shy, apologetic smile. But, in fact, I was the one who needed to apologise for my impolite stubborness.

34. Queen Mary's Peak

Saturday, October 16th. The wind had been howling around the house for three days, so I decided tro stay in bed with my library book. But then I went to the bathroom, and through the little window I looked up at the rim of the Base, sharply demarcated against a deep blue sky. Wow! I had to be up there! It was already 10 o'clock. I made a quick round through the village to see if I could find a guide for the day, but there was nobody available, as usual. So I quickly assembled some sandwiches, threw a water bottle in my backpack, and went to Pigbite. It was already past noon when I reached the Base. It had not yet crossed my mind to go all the way to the Peak, I just wanted a Moorhen day on the Base. But there it was, the Peak, cloudless, and looking so near you could almost grab it. Unimaginable that it was still 1300 metres further up and five kilometres away. Irresistible, so I changed my plan. This would be my only opportunity to get up there. It was crazy to go to the Peak so late. The sensible thing would have been to start at six am. I guessed I would need about three hours to get up there, and I knew I needed to be very strict with myself about descending in time. I could not risk negotiating the Pigbite trail after sunset. Supposing I stayed up there till half past three, then I could be back at Big Green hill two hours later, which would give me plenty of time to get down to the Settlement in daylight.

The route up, from Big Green Hill, looked pretty straightforward. I regularly turned around to look at the lay of the land and to take a compass bearing at Big Green Hill. As an extra precaution I did not cross any gulch, carefully staying between the same two spokes of the wheel. That would guarantee that I could get down safely, even if rain changed the gulches into torrents.

The colours were magnificent. First the deep green and brown shades of the fern bush, then the yellowish brown tints of the grass zone, and above that grey, black, and purple colours of the bare volcanic rock. High up some gullies were filled with snow and the Peak itself was lightly powdered with frost. The sky was still cloudless but above the sea, lower than where I was, there was a cloud layer, just about Base level. There was a strong wind from the south, later in the day turning towards the west. I was in the cloudless triangle in the lee, just like on the day of arrival. So, if the wind turned further, I might have to deal with clouds and fog on my way back to Big Green Hill.

Above the grass belt, at about 1500 m. altitude, the vegetation changed into alpine tundra, many plants growing in cushion shapes, here and there speckled red with the berries of *Nertera depressa*, the tasteless berries which islanders use to decorate their pies. To my surprise I met a Starchy in this bleak environment. At 1800 m. altitude, the vegetation became sparse, and shady rockwalls carried curtains of icicles. All stones had a small icecap on the windward side formed by compacted snow, frozen during the night.

I thought I would easily meet my schedule, until I discovered that the Peak I had been looking at all the time wasn't the true Peak yet, so it was already half past four when I reached the real summit. The last bit was very steep with the notorious loose cinders: three steps up, two steps gliding down, very tiring. No time left to stay up there, I had to go back down almost immediately. It wasn't very agreeable up there either with a fierce feezing wind. The crater lake was partly filled with ice floes. With all the iced stones the Peak looked as if sugar had been strewn all over the place.

The view from the Peak was limited to the Peak itself and the sea of clouds all around. On the south side of the Mountain the clouds were chased uphill in strange whirls. In order to look down on the island, you had to get a little away from the summit. On the east side I looked into some beautiful craters between Sandy Point and Stony Hill, some of which were filled with clear blue lakes which are not shown on the map. To the north, I had a magnificent view all across the whole area between Hottentot Gulch and the Ponds, in the middle of which was this little green wart, Big Green Hill, which did not look big at all from here.

The world around the summit is barren and desolate. Not a sign of a bird. In the old days the great Tristan Albatrosses nested here in the cold, but they have all been eaten. Inaccessible still has one or two pairs, just like sixty years ago when the Norwegians were here. Perhaps even the same individuals.

The descent was problem-free. I watched my footing extra carefully, well aware that the likelihood of spraining an ankle is much greater going down than going up. At six pm I went down the Pigbite Trail, getting home in time to have a shower before Noe and I went to have dinner with the girls next-door. A glass of wine made me really sleepy. I told my story. They all thought I had been naughty, or even crazy, but they were also a bit jealous. None of them had been any further than the Potato Patches. They were all promised to be taken to Burntwood or the Ponds, or on a boat trip around the island, but for reasons we now know all too well, this never happened.

Two days later Herbert went to the Peak with the Whale People. They knew about my escape but Herbert didn't. On the other side of the crater lake they saw my footsteps in the snow. Yetis on Tristan? Herbert wanted to have a closer look but the Whale People suggested going back on the same side, so they left it. Probably a stray sheep they decided.

Of course the islanders are absolutely right to be so careful with guests. If I had hurt myself up there I could have been in serious trouble. Therefore I took the greatest possible care during walking and took my other precautions. But, on the whole, I did not find the ascent very difficult, apart from the last bit being tiring. In fact, I only needed my hands in the steep part of the Pigbite Trail. From Big Green Hill up it could basically be done with your hands in your pockets.

Thanks to the islanders' care, accidents on the Mountain are very rare and guests have never been lost. The English boys of the Gough Island expedition were allowed to make a two-day excursion to Soggy Plain without a guide, but when they were cooking their meal in front of their tents, near Third Gulch, an islander suddenly appeared to check on them. The same boys, all experienced mountaineers, tried to climb the rock wall above the Settlement, against advice. In the last few metres below the edge they got hopelessly stuck. Every rock they tried for a hold or a step broke loose and tumbled down. They could not move forward or backward. In the village people were looking up at them in tears, convinced they were lost. But a few islanders managed to get above them with ropes and saved them. Incidents like that are not very helpful in allowing guests much freedom of movement.

A few months after my visit to Tristan, the Admin went for a lone walk at Pigbite and fell, badly hurting his leg. He was not found until next morning, suffering from hypothermia and blood loss. A narrow escape. So the islanders are absolutely right not to allow you to go up on your own and we should all abide by their rules. I have to apologise for having been utterly irresponsible. Yet I never regtretted what I did. In Europe I have climbed several summits in the Alps and the Pyrenees, but I never felt so much on top of the world as on Tristan's Queen Mary's Peak.

35. The Molly study plot

My second visit to Sandy Point, to check my albatross nests, never happened. That was not really Jeff's fault but the weather played tricks. After some days with terrible weather, big waves crashing into the harbour, time became a problem, and we might risk not being able to get back in time for the *Agulhas* on her return voyage. Jeff promised me to do the check in a month or so, and I paid him in advance for the days he would need to do that. I explained to him that he needed to note the fate of each nest exactly by number, and that the check had to take place before the first chick hatched. Otherwise you could not distinguish between lost eggs or lost chicks and this study was purely about eggs.

The only thing still to do was a visit to the Molly Study Plot with either Connie or Ian. Some ten years ago, the Molly Study Plot, on the lower Base just above Hottentot Gulch, was established by an expat teacher who liked to take schoolchildren up there. There are about twenty Molly nests, which have been neatly mapped. Nowadays, teaching is done by Tristanian women who do not go up there with children. So monitoring the birds is now the responsibility of the two ornithologists of the island, Connie and Ian, supervised from a distance by John Cooper in Cape Town. John had asked me to take a look, because in the previous year, when Connie was in England for police training, he had not received any results. So I asked Ian if I could see his data from the year before. That was fine, but every time I asked, he had left his notebook at home. Finally, on my last day on the island, I was able to get his notes from his father. The plot had been checked last year, but only very late in the season. Ian said he had not sent the data to John because he still wanted to go over them with Connie and add details from his memory, a year later.

The problem with a late start is that you never know

how many nests have been lost already before you got there. This is not just a Tristan problem, but a serious matter everywhere. If you find a number of nests of a particular species, and later you find that a certain percentage of these nests hatched, and you call that the hatching success, your estimate is always too high, which is not easy to understand. It may sound trivial, but the problem is that nests which were lost before you found them are excluded from the sample. I illustrate this with a hypothetical bird species, which hides its nests so well, that we are not able to find them. But there is a predator which smells them and removes more than half of the clutches, without us knowing about it. On the last day before hatching our bird shows a conspicuous behaviour which allows us to find them, at last. Almost all these nests will hatch the next day, so we will find an apparent hatching succes of almost a hundred percent. In this example it it obvious that this conclusion is wrong, but in real life this is also the case. Apparent hatching success, expressed as the percentage of nests found successfully hatching, is always too high. Always. The only way to tackle this problem is to calculate the probability that a nest survives one day, and use that probability to estimate the final success. To apply this method you need to know exactly when each nest was found, and when they were checked, because you need those intervals for your calculations. In this way we have calculated nesting success for thousands of nests of Lapwings and Black-tailed Godwits, to study the impact of predation, grazing, and grassland management.

This year, neither Connie, nor Ian had yet been to the Molly Plot, even though the birds had had eggs for more than two weeks, perhaps even three. So I kept nagging at their doorsteps about going up soon, and I also tried to explain to them why this was so important, but they probably found me a bit of a nuisance and they never had time to go there with me.

I often had a cup of tea or a beer in Connie's kitchen.

We talked about penguin censuses, and about Nightingale and Inaccessible. What a shame that I could not stay any longer because next month they would go there. And of course we talked about the Molly Plot. Every time we agreed to go up, but every time it had to be postponed because of important police work, sheep shearing, or planting potatoes. His house (masonite walls, glossy paint) was still decorated with Welcome Home signs, to celebrate his return from England, and in the kitchen there were still tons of leftovers from all the cakes and pies that were brought. So on every visit I was given a generous piece of something. I tried to explain again why it was so important to go there early in the season, but Connie said there was no need. After all these years, he knew the birds and their habits so well that he could reconstruct everything that happened up there, even from one late visit.

When I argued that data collected in such a way would be rather useless, Connie exploded. Now I had done it again, alienating an islander, stubbornly insisting on knowing better, Mr know-it-all with all my useless diplomas. Connie exclaimed that he did not want to be told how to do his job by any learned outsider. I had no idea what was going on up there, did not know the animals, or the island in general, and could I please leave the house. On legendary hospitable Tristan, with all the doors always open, I was kicked out! I really had gone too far. I stammered that I did not mean it so seriously, and had hoped he would appreciate my assistance and apologised. My apology was accepted, but I had to leave anyway.

My chances for guided tours on the mountain evaporated. It was too late to go back to Sandy Point and Jeff was nowhere to be found. And now I had ruined the possibility of visiting the Molly Plot. I could only sit in the Settlement, waiting for the *Agulhas* to return.

Of course I escaped to Big Green Hill and the Ponds once more, on a day with fine weather. This time I explored the area close to the cliff edge, with lots of beautiful. densely vegetated gullies. Moorhens were everywhere, and I even found families with small chicks. I never saw full grown brown juveniles like the ones Michael saw in April. In the families with chicks the parent would make soft chuckling noises, the chicks answered with squeaking sounds. For the first time I saw a Moorhen walking around an incubating albatross. The two birds showed no interest in each other. The Moorhen scratched the soil as if looking for woodlice or snails.

Suddenly I was startled by a loud bang from below, reverberating against the cliffs. All around the Moorhens called "chack-chack", and a passing Skua made a sudden twitch in flight as if it had been hit by a hail shot. A large cloud of dust drifted away from the factory, so I thought there had been a serious accident. I scanned the village with my binoculars, but could not detect any panic. No ambulances with blue flaslights, no-one running, no reaction at all. A few men in blue overalls were standing still, looking at the factory with their hands in their pockets.

In front of the factory a large obsolete crane stood on a concrete block. It spoiled the ocean view from the director's house, and every time he came back from Cape Town he hoped it might have been removed. Now he saw an opportunity: the harbour project. Tristan never had a decent harbour. First the boats just landed on the beach, which is now covered with thick lava from the new volcano. But between the lava and the shore a natural harbour had formed, allowing small boats to enter safely. Successive gales have eaten away the lava and this natural harbour soon desintegrated. Expat technicians started to build a new harbour a bit further to the west, The islanders said it was not a good place because of a number of reefs, which could be passed most of the time, but could be a problem at low spring tide. The technicians did not listen and went on with their job and, indeed, the reefs between the piers are a problem today. In addition the concrete walls had been severely eroded by wave action, so the whole harbour needed a thorough renovation (which was the reason for my delayed visit, because all potential guides were employed in the harbour project). The notorious reefs were attacked with dynamite. The director of the factory asked the harbour technicians to blow up the crane for him, and so they did. The problem was that these boys were used to underwater explosions and had slightly underestimated the effect of dynamite in open air. The crane was indeed gone, but the facing side of the factory was seriously damaged too. Corrugated iron was folded like paper and it looked like a war zone. Windows were shattered all over the village and shrapnel from the factory was found as far away as the school. It is a miracle nobody was hurt.

36. Theories

Whether we believe that the Island Cocks on Tristan were brought here by Michael Swales, or could be survivors from the original stock, either way it is hard to understand why they were so succesful in recolonising the island while were unable to survive earlier. I have a theory that the rats may have helped Island Cocks to survive, instead of wiping them out.

In previous centuries the Moorhens had a hard time. First came people, who hunted them down with their dogs, but what was much worse was the establishment of feral populations of cats and dogs which roamed all over the island, able to go where man hardly ever ventured. Dogs and cats could never survive on Moorhens alone but will have killed them whenever they found one. Their staple food was no doubt the unlimited supply of fat seabird chicks, which could be found throughout the year, some species nesting in summer, others in winter. Seabird populations must have started to dwindle, which also affected the scavenging Moorhens, who depended on seabird colonies. This situation, heavy predation plus deprivation of food resources, may explain why Moorhens were already close to extinction before the first rat turned up in 1882.

Popular thought is that the rats gave them the final blow, but the story may be more complicated. Undoubtedly the fast multiplying rats depleted the last colonies of the smaller seabirds, reducing the survival chances for cats and dogs, but also for themselves. Perhaps the rat population plummeted because of starvation, which would explain their extraordinary hungry behaviour, digging up potatoes, and climbing the trees for peaches, causing an economic crisis for the people of Tristan da Cunha. Cats and dogs died out spontaneously, quite miraculously. This process may have been greatly helped by the rats. With lower population densities of rats there was room for some recovery in seabird species. This may lead to a new boom in the rat populations but eventually rats and seabirds will reach a dynamic equilibrium, in which there would also be survival opportunities for cats and dogs. Cats are no longer kept as pets on Tristan so they cannot run wild anymore. Fine, keep it that way! After the volcano and the evacuation of the island a few dogs ran wild but they were found and shot.

In the absence of feral cats and dogs, the lower densities of rats, and perhaps slightly recovered seabirds, there may be better chances for Moorhens to survive. The increase in Molly numbers, thanks to protection, also helps, as parents feeding their chicks often spill food which is eagerly picked up by Moorhens. And who knows, they may take the egg every now and then...

Apart from scavenging Moorhens can feed on invertebrates, which have become much more common now than in the past. There used to be no centipedes and woodlice on the island, but these are very numerous now. The brown, rotting masses around the stems of tree ferns teem with woodlice, and I have seen Moorhens scratching for food like a chicken.

Speaking of wild theories, I would finally like to turn my attention to the relationship between Moorhens and albatrosses. As pedantic, know-it-all outsiders, we may too easily dismiss stories of egg-snatching Island Cocks, classifying them as island folklore. Don't forget that Island Cocks in the nineteenth century had the same bad reputation as the ones today. People nowadays don't believe there ever were Island Cocks on the island, so this knowledge has not been passed down from father to son, but must have been acquired independently, indicating some truth.

During egg laying, birds need an extra calcium supply to produce egg shells. In an environment poor in calcium this may be a problem. In our forests birds like the Great Tit get most of their calcium from snail shells. When acid rain hit our forests on poor sandy soil the snails disappeared as they could no longer build their shells. The Great Tits, in turn, encountered grave problems in their egg-shell production. They cleverly plundered rubbish bins near picnic tables, searching for discarded shells of hard-boiled chicken eggs. Others pecked away bits of mortar between bricks, especially in old, withered walls.

Tristan is by origin poor in calcium, consisting of volcanic rock, covered with acid bog vegetations. Seabirds obtain the necessary calcium from marine organisms, and through their droppings, lost eggs, and dead chicks, they supply calcium for the land birds. Starchies are also scavengers, like Island Cocks. Perhaps the collapse of seabird populations, caused by cats, dogs and rats caused problems for the land birds in building up their eggs. Maybe this explains why the thrushes are so rare on Tristan.

This also makes it quite understandable that Moorhens are interested in albatross eggs. They will certainly take abandoned eggs. Michael Swales caught Moorhens using egg shells as bait, and Mike Richardson found egg-shell fragments in stomachs. Gough Moorhens are also known to take eggs. But, whether the Island Cocks really plunder albatross nests and take the egg while the incubating bird is sitting on it, remains a question, as nobody has actually seen this happen. And I (pedantic outsider) find it hard to imagine that they really would pose a problem at population level. But strange things happen in nature.

The Turnstone is a small wader species, breeding in Scandinavia, Siberia, and northern North America. In winter we can see them along our coastlines. Many migrating waders frequent our tidal mudflats but turnstones prefer rocky shores. We can see them turning pieces of seaweed or pebbles, in search of food items, hence the name. In West African fishing villages we see them scavenging around fishing boats, and even on rubbish heaps in the centre of the village. They also feed on the carcasses of sharks which are left to rot on the beach after the fins have been sold to Chinese merchants. On one small Finnish island, Turnstones discovered that the eggs of the Common Gull are delicious. They just walk around unobtrusively, through the colony, and when they see an unguarded egg, for instance when the owner is involved in a squabble with its neighbour, they run towards it and peck a hole in the side to get at the contents. True Island Cock behaviour. The gulls pay no attention to the Turnstones. They have not yet learned to recognise them as an enemy.

This is reminiscent of the feeding behaviour of Sheathbills in Antarctic penguin colonies. Like the Finnish Turnstones, they walk around in the colony looking for dead chicks or unguarded eggs. The penguins don't pay any attention to them, while their reaction to approaching Skuas is very fierce and aggressive. This may indicate that the Island Cock behaviour of Sheathbills is a relatively recent development, speaking on an evolutionary time scale. The Lesser Sheathbills of the Crozet Islands in the southern Indian Ocean, discovered another nice trick. They approach penguins which are just about to regurgitate food for their chicks. The Sheathbill then flies right into the penguin's face, with the result that the food drops on the ground instead of inside the throat of the chick.

These examples show that it is quite possible that Island Cocks learn to utilise albatross eggs as a food source. Only the story of the two co-operating partners (which nobody has actually seen) goes a little bit too far in my opinion. It reminds me of a similar piece of island folklore on Schiermonnikoog, my favourite little Waddensea island, where I studied the breeding biology of the Shelduck for a whole year. Shelducks breed in rabbit burrows in the dunes, on the North Sea side of the island, but have to guide their chicks to the mudflats of the Waddensea on the other side, a few kilometres away. People on the island believe that these mean Herring Gulls ruthlessly kill all the chicks when Shelduck families make this dangerous trek. Yet, outside (pedantic) ornithologists see that the population is not decreasing. According to the islanders the gulls have become so clever that they no longer only attack families on the road, but also wait for the piping sound of hatching eggs, listening at the entrance of the burrow. Finding eyewitnesses on Schiermonnikoog will no doubt go around in a circle, just like on Tristan da Cunha.

37. Going home

The *Agulhas* was back. The helicopter deposited a few crew members who wished to plunder the post office, and Mike Frazer, who had a wonderful time on Gough. Mike and I walked to the Potato Patches to meet his old friend Nelson Green who was at work there. They talked about their shared past on Inaccessible in 1983 and Nelson's stay in Cape town last year where he went for an operation. Mike often visited him in the hospital.

We wanted to take the bus back to the Settlement but it was almost full, and could only take Nelson. Mike and I had to walk but Nelson would let Winnie know that we were coming for lunch. When we arrived at their house the table was alreay full will lamb dishes, peas, beetroot, mashed potatoes, and of course, Tristan pastries. Winnie had made a 'spotted dick', which sounds like a horrible contagious disease, but was actually a pudding, made of flour and raisins wrapped in a cloth and boiled in water.

In the late afternoon, there was a farewell reception at the Admin's house because he was leaving for three months. Walking around with a beer I felt hopelessly superfluous. I had nothing to tell the islanders anymore, and I felt they had nothing to tell me either. People looking in my direction were not looking at me but through me. My falling out with Connie Glass still bothered me and I felt my popularity had reached an absolute low. Mr know-it-all, sympathising with those despicable Island Cocks. And suddenly I had seen enough of them too, with their all-important sheep and potatoes, and their "we is used to it," implicitly meaning "you is not", and their perpetual "maybe tomorrow."

That night, Noe was having his own farewell party in the Prince Philip Hall, complete with disco and the Tristan Rock Lobsters and free drinks for everybody. His last days had been very busy, taking all the blood samples, and getting them, as fresh as possible, in the fridge on board the *Agulhas*. I was not in the mood for a noisy disco night. Eyedoctor Liz and dentist Charlotte too had no enthusiasm to party. They feared that free drinks would result in everybody getting terribly drunk. Charlotte had never seen so many cavities in her life and had to work overtime to fill them all. The myth of the impeccable teeth of Tristan is history. The girls took the last helicopter flight back to the ship, and when they asked me to join them I could not resist, liking spectacular helicopter flights too much. The evening was beautiful. We repositioned to the lee near Sandy Point, with thousands of albatrosses and petrels returning to their colonies in beautiful sunset light. A much better good-bye than a disco night.

Next morning the other passengers were craned on board in the bird cage. Several islanders came on the ship to say good-bye, including Jeff. He emphasised that during my whole stay there had been no day suitable for going on the mountain. And I would have been physically unfit to reach the peak anyway, but he looked at me in a strange, suspicious way. Finally, after a minute of silence, he slapped me on the shoulder, and said "Take care!"

It took hours for Tristan to disappear. The island is too high to sink below the horizon, but instead it slowly dissolved. With a sailing albatross in front of the silhouette it became the classic picture, a symbol, like in the passport stamp. The dream island. You can ask yourself whether you should visit dream destinations or if it is better to leave them as dreams. Mythical places, like Timbukto or Cape Horn, tend to become 'normal' if you actuallly visit them. And the village of Tristan da Cunha also turned out to be extraordinarily ordinary. You can spend your whole life dreaming about an island, but when you finally get there, the dream is over. Call it the 'de-mythologising' of a boy's dream. Yet, I would never use this as an argument to stay at home.

I discussed my decision to leave the island prematurely (did I really flee?) with Noe, who was disappointed I did not join his party, which went very well. Nobody got really drunk and it had ended with the traditional pillow dance. I certainly regretted missing that. But our experiences on Tristan had been completely different. His project was an overwhelming success, thanks to the endless co-operation of the islanders. He came close to leaving behind the proverbial 300 best friends. My albatross project, on the other hand, had been a total failure. My results on the biology of the Moorhens were also negligible. And I did not manage to have a look at the Molly Plot project, as I had promised John Cooper, coming up with absolutely nothing (John was not surprised). Fortunately they were all private arrangements so I had no responsibility to answer to funding agencies. That would have been a disaster. In contrast to Noe's experiences, the little I managed to do, including my illegal excursions on the mountain, I achieved not thanks to the islanders, but rather in spite of them. I am sorry if this sounds harsh. However, in contrast, I left with memories of the most remarkable trip I had ever made in my life. I would not hesitate to return!

In hindsight, I have been the victim of my own naivity. Thinking that I could solve the problem of finding a guide by simply hiring someone for every day of my stay was utterly naive. With all I have read about Tristan I should have realised that this is not how things work. Over a period of three weeks I only saw Jeff on six days. Four days Sandy Point and two days the Caves. On the other days nobody else was willing to guide me anywhere because Jeff had the official contract and they did not want to step on his turf, with the exception of Ian Lavarello, Conrad Glass and Clive Glass. Ian and Conrad promised to take me to the Molly Plot and Clive wanted to go Island Cock hunting around the Ponds with me, but they all kept saying "Maybe tomorrow" until I left the island. Could I have solved the problem in another way? I don't know. Perhaps the whole project, as I had designed it, was impossible to execute under the prevailing conditions on

Tristan. I think that today it would be easier. With the availability of the RIB *Wave Dancer* it would be possible to do the fieldwork at Sandy Point on day trips from the Settlement, but I don't think anyone will ever pick this up again.

Eight months later, to my surprise, I received a letter from Jeff, telling me that he went back to Sandy Point and found that one third of the albatross eggs had been lost, either to Island Cocks, or to Skuas. Unfortunately, without exact figures, this data was of little use. But I appreciated it immensely that he actually went back and had not maybetomorrowed it until after the breeding season.

Looking back, my frustrations have become an integral part of my extraodinary experiences on Tristan da Cunha. And my scientific failures? Frankly I don't care. I had a wonderful time.

Part 3. Tristan again

38. Atlantic Odyssey

My book on the Moorhen of Tristan da Cunha was published in 1997. For the occasion, I organised a mini-symposium about Tristan for friends and family. I showed a selection of my own slides and explained the riddle of the Moorhen. I invited captain Albert Veldkamp to show some of his slides of his visit with the RMS *Saint Helena* in 1993. Marietje Brander, the 82-year-old daughter of Tristan historian Jan Brander, told the story of the wooden shoes her father collected 60 years earlier. She brought the pair of clogs her father kept for her and she asked me to take these to Tristan on the first possible occasion (I already knew I was going back the next year). Janneke told the story of her near-shipwreck with the *Bris*, and radio reporter Marnix Koolhaas let us listen to some of his sound fragments which he had recorded just the year before. He managed to get the famous sentence "there is no news today" on tape. For snacks I served Tristan Potato Cakes, following the recipe from Pam Lavarello's cook book. I had no petrel fat so I fried them in oil.

A few days earlier I was interviewed by a national newspaper and they used the headline "BIOLOGIST ORGANISES SYMPOSIUM ABOUT TRISTAN DA CUNHA". Hell broke loose. The article mentioned my home town, so people phoned the library, the tourist office, the police, the book shop, and God knows who else. But fortunately, nobody knew where or exactly when it was happening as it was a private thing. Nevertheless, some people managed to find out so we had a few uninvited guests. An elderly lady walked in, right in the middle of my own presentation, a big folder under her arm, filled with newspaper clippings. She walked right up to me, but her son grabbed her by the tail feathers to keep her in check, loudly hissing: "Mother! Not now!". There was a man from Katwijk, Pieter Groen's hometown, who said he was a distant relative of Pieter. He brought one of

263

the labels that were attached to the sacks filled with clogs, addressed to the reverend Harold Wilde. He wanted me to take it to Tristan with Marietje's clogs. And there was a banker who managed an offshore RABO bank office in the Channel Islands. He was intrigued by Tristan, as some years ago they had tried to establish an offshore branch at Tristan da Cunha. Imagine the drug barons hiding their millions on Tristan. Great opportunities. Well, maybe tomorrow.

In the 1970's and 80's, the Arctic Centre of the University of Groningen ran an archeological project in Smeerenburg (Smear city, Filth town), on the arctic Island of Amsterdam in the northwest of Spitsbergen (not to be confused with Amsterdam Island in the Indian ocean), where the Dutch had a very successful whaling industry in the first half of the 17th century, following the discoveries of Willem Barentsz. In less than a century we managed to bring the whales so close to extinction that the business had to be abandoned. In order to conduct their research the scientists from Groningen needed a ship. The University obtained a governmental vessel, the *Pollux*. It was renamed *Plancius*, in honour of the Reverend Plancius, who was the genius behind the expeditions of Willem Barentsz. To facilitate operations with the ship the Plancius Foundation was established. After the Smeerenburg project finished the foundation continued. They began to take passengers and gave logistic support to various expeditions. The passenger business was very low key. They had to make their own beds and assist in the kitchen peeling potatoes. For many years Albert Veldkamp was the captain of the ship. Gradually, the business expanded, and the board of the foundation felt that they should try to find more ships and also move into the Antarctic tourist industry. Meanwhile, they had merged with Oceanwide Expeditions, which, at that time, only had one ship, the *Rembrandt*, sailing in the Carribean. Their big opportunity came in 1991, when the Soviet Union collapsed, and all sorts of research vessels (spy vessels if you like) became available. The first Russian ship they were able to lease was the *Professor Molchanov*. They shared it with another tour operator, Quark Expeditions. Quark would have it in the Antarctic Season, Oceanwide in the Artcic. Later, Oceanwide managed to get their hands on three more Russian ships, *Multanovski*, *Mikeev*, and *Maryshev*, all professors too. These ships could have around 50 passengers on what is called expedition cruises. No ball rooms, no swimming pools, no captain's dinners, no ties and jackets. Instead, a serious lecture programme and landings by zodiac, with rubber boots, between penguins and seals. With those four ships they developed their own Antarctic programme. All ships had to sail back and forth, between the Arctic and the Antarctic, with the seasons. That is where the idea of the Atlantic Odyssey was born. Oceanwide calculated that enough passengers could be found to fill one of the four ships to make the repositioning from south to north a profitable cruise (interestingly enough, the reverse trip, from north to south, never became an option. People did not seem interested and also, in the Southern Ocean, you would encounter serious contrary winds and currents). The choice was the *Professor Molchanov*. The first Atlantic Odyssey was going to take place in 1998 in March-April (southern autumn, northern spring). And it was quite obvious that I would be on board as a lecturer. On this voyage, I took Marietje Brander's wooden shoes to Tristan and donated them to the museum on her behalf. One of our passengers was ex-whaler Albert Veldkamp. When we reached the Southern Ocean he was on the bridge, exclaiming: "Where are the whales?", as we saw none. We said: "Well, you bloody well shot them all!", and he said: "No, no. When we quit, there were plenty left." We did not argue the matter and let it rest.

I was a bit wary about meeting Connie Glass again on Tristan, but when he saw me he looked genuinely pleased to see me. When he came on board with the immigration officers, all in impeccable uniforms, and went to the captain's cabin to stamp our passports I was invited to join them. When we went to Nightingale Lewis Glass was one of our guides. He too seemed pleased to see me. "Here is the man who thinks an albatross drops dead if you throw it off a cliff!" he said, immediately putting me back in the place where I belong: the ignorant outsider.

My wife Dineke did not dare to come on the first Odyssey, being too afraid of seasickness, but I convinced her to join me on the next one. Standing together on Nightingale, hand in hand, at the spot where we would have erected our hut thirty years earlier, was kind of magic. On Tristan people were really proud that I not only came back for a third time, but this time even brought my wife to show her the place.

In 2010, the four Russian ships were replaced by two Dutch ships, the *Plancius*, copying the name of the other *Plancius*, which sailed around Spitsbergen, and the *Ortelius*, a powerful ship able to negotiate a lot of ice. Both have room for over a hundred passengers. The reception of the 'New Plancius' is adorned with a painting of the 'Old Plancius', painted by Albert Veldkamp.

My lectures on board were about the Moorhens of Tristan da Cunha, the history of Tristan and all other islands we visited, and Antarctic exploration, but also about waves and tides, ice and glaciers, sea currents and atmospheric circulations, and a few other topics. In addition, I had the habit of drawing a daily cartoon on the whiteboard secretly at night. Sometimes, in the morning I discovered that my cartoon was wiped out. That was annoying, of course, and it took a while to find out that that mostly happened when I depicted a ship in distress (my best one being on April 1st, a sinking ship, everybody in the life boats, except one face behind a porthole, saying "April Fool, ha! They won't fool me!"). Sailors are known to be superstitious, especially the Russians, and they found such cartoons threatening. Likewise, you may not whistle a happy tune on any ship. Whistling invites the wind to increase to gale force. I was even instructed by the captain not to draw such things anymore. I also had a popular sequel on Sperm Whales. Number one: tadpoles jumping on the horizon. Comment: "Look, that must be baby Sperm Whales." Number two: a breaching Sperm Whale with a condom over his enormous head. Comment: "Look, that Sperm Whale is a Safe Whale." Number three: a very angry looking Sperm Whale right below the keel of the ship, bubbling: "One more dirty joke about Sperm Whales, and I will give them a blow job they'll never forget!"

Apart from our lectures there would always be a daily recap in the late afternoon, just before the bar opened for pre-dinner drinks. We would go over the things that had happened that day, and the birds and mammals we had seen, often higlighted with a five-minute mini-lecture. My contribution was usually a short story from history, often classified as a 'gruesome story', of which I had quite a collection.

The Atlantic Odyssey (AO) starts in Ushuaia, the capital of the Argentinian half of Tierra del Fuego (the western half is Chilean), on the northern shore of the Beagle Channel. The landing at Ushuaia airport is quite spectacular, swooping down between the mountains either side of the channel. Coming from Europe you have to change airports in Buenos Aires, and once, arriving at the domestic airport, a strike was announced minutes before our flight to Ushuaia was due to leave. We had to stay overnight in the city, and arrived too late in Ushuaia. Fortunately, the expedition leader was with us so the ship waited. Otherwise it would have left without me and a few passengers. Since then, I always took an extra day in Ushuaia, to be sure. An extra day there is no punishment. Along the shoreline you see Kelp Geese, Chilean Skuas, a local brand of Night Heron, wintering White-rumped Sandpipers from North America, and Flightless Steamer Ducks. In the pastures there are Chilean Lapwings and Upland Geese, and in the *Notophagus* (southern beech) forests, you can find the spectacular Woody Woodpecker-like Magellanic Woodpecker, and even tropical looking parakeets and hummingbirds. Ushuaia is a friendly town, which I have seen grow over the years, now sprawling all over the mountain side. At the same time, I saw the Martial Glacier, which can be seen above the city and can be reached on a nice day hike, shrink with an astonishing speed. Soon it will be gone completely. Further west on the Beagle Channel, with a lot more precipitation, large glaciers still flow down to sea level at the end of deep fjords, but they too are retreating at an alarming rate.

Ushuaia has nice shops and good restaurants, where I eat my favourite roasted lamb with a good glass of malbec. In some restaurants you can watch the lambs being roasted, whole lambs, skinned, gutted and decapitated, spreadeagled, crucified to racks, arranged around a huge fire. Every now and then, the sweaty cook would enter with an enormous knife, to chop off chunks that are done. You pay a fixed price and can eat as much as you want. If you want your lamb not overcooked but juicy, you order 'muy jugoso'.

On the opposite side of the east west running Beagle Channel, lies Isla Navarino, a large, mountanous island, Chilean territory, part of a large archipelago of islands, all the way south to fabled Cape Horn, where so many ships have been lost in unbelievably strong gales. The island of Cape Horn has a manned lighthouse. The Chilean flag there has to be replaced almost annually, as it is blown to shreds all the time. I have landed on Cape Horn, and was surprised to find small songbirds like wrens and thrushes in such a bleak and windy place. From the lighthouse you can walk to the famous albatross monument, a large structure that can be seen from a ship at quite a distance. You are advised to stay on the boardwalk through the bog, because of landmines that were put there by the Chileans to deter the Argentinians when they were fighting over three insignificant islands in the eastern entrance of the Beagle Channel: Isla Nuevo, Picton and Lennox. It all had to do with establishing the starting point of the line that runs down to the South Pole and which defines the territorial claim there. The Chilean and Argentinian claims greatly overlap, not just with each other, but also with the British claim. Although internationally, none of these claims is recognised, Britain still thinks that the Antarctic Peninsula is British territory. For the Argentinians it is Argentinian territory, for the Chileans it is Chilean. This issue can never be solved in a peaceful way and it is my conviction that this problem is safeguarding Antarctica. I am sure, had there been no such issues, the claiming countries would have divided the cake a long time ago.

The conflict over Nuevo, Picton and Lennox did not escalate into a real war. The Pope was asked to mediate, and he decided that the islands were Chilean. They are uninhabited, but there are some sheep, and Chilean landmines make them fly occasianally, as the Argentinean mines do in the Falklands. Speaking of the Falklands, in 2007 I was amazed to see how Ushuaia was celebrating the 25th anniversary of the liberation of the Malvinas. The whole city was decorated, and there was a large photo exhibition in a park near the harbour. Not a word about the British taking the islands back, very strange. In the same park there is a large monument commemorating the war, with the ominous words 'we will be back' (volveremos). At the same time Ushuaia was officially appointed to be the capital of the Malvinas too. Most Argentinians probably don't realise that the name Malvinas is just as un-Spanish as Falklands. The islands were named Iles Malouines by French explorers, after the Fench port of Saint Malo.

We leave Ushuaia around six pm, after provisions have been loaded, passports have been stamped and the ship is cleared. We sail east on the Beagle Channel, with spectacular views. It is striking how low the tree line is, and how abrupt the transition to alpine tundra. Above that, snowy peaks, Tierra del Fuego on the port side, Isla Navarino on starbord. Around the ship plentiful Black-browed Albatrosses, Sooty Shearwaters and occasionally

Magellanic Penguins. We have a pilot on board until we reach the disputed islands, where the channel branches. Around 3 am we reach the open ocean and are exposed to oceanic swells which means that not many passengers appear at their first breakfast on board.

It takes about two days to reach the Antarctic Peninsula, where we always try to make a continental landing, usually near Brown Bluff, in Hope Bay, near the tip of the peninsula. We also sometimes visit the Argentinian base Esperanza there (esperanza = hope), where the Argentinians make a point of this not being a base, but an ordinary Argentinian village, with families with children, a hospital, a school and a church. Pregnant women have been taken there to deliver true Antarctic citizens.

During the crossing the birdwachters are very keen to distinguish the various species of petrels and albatrosses. There are Black-browed Albatrosses, Grey-headed Albatrosses, Light-mantled Sooty Albatrosses, and the famous Wandering Albatross, the largest flying bird on earth. The birdwatchers are especially alert when these giants approach the ship, which they often do, because at close range they may turn out to be the Royal Albatross, a lookalike all the way from New Zealand. Like many other species, the Royals have been split into two separate species, hardly distinguishable, the Northern Royal, and the Southern Royal Albatross. Both can be seen in the waters south of Cape Horn.

A significant number of the passengers are birdwatchers. They are often British, booking through the British birding travel agent Wildwings. Most of them are male, young, and they all hope to see the Flighless Rail of Inaccessible. The other passengers are usually much older, often pensioners, of mixed gender. There are sometimes tensions between the birders, who always elbow their way to the best position at the stern, and the non-birders, and it can be quite a challenge to keep everybody happy. Most passengers are British or Dutch, but we have all sorts of nationalities on board. And we have the craziest kind of people. Another significant group are the people who collect islands, countries, or territories, like the members of the Travelers' Century Club (TCC), founded in 1954. To be a member you have to have visited more than 100 countries or territories. They maintain a list of territories, which has to be updated on a regular basis, as borders change and countries split up into independent parts. On their most recent list (2020) they distinguish 329 territories in the world. Then there is the Extreme Travelers Club of the Most Traveled People (MTP). They too have a list, which is much longer. They distinguish 995 different travel destinations, 259 in Europe alone. A difference is that for instance, for the TCC, The Falklands and the Falkland Dependencies (South Georgia, South Orkneys, South Shetlands, and the South Sandwich Islands) are counted as one territory. Likewise, Saint Helena, including Tristan da Cunha and Ascension, is one territory. For the MTP, these are all seperate destinations. Then there are stamp collecters, who are only interested in the post offices of the places we visit. And we had a world cyclist on board. He had cycled from Ireland to Vladivostok, from the North Cape to Cape Town, and from northern Alaska to Ushuaia, but he had no interesting stories to tell. He was usually sitting alone, arms folded around himself. Perhaps he had a form of autism. On his account, I once decided to compose a limerick instead of a cartoon:

There once was a man on a bicycle,
Who wanted to ride on an icicle.
It did not go well,
He slipped, and he fell,
So now he is riding a tricycle.

After the Antarctic Peninsula we head northeast. In earlier versions of the AO we would visit other Antarctic islands, such as Elephant Island (like Tristan, a destination that I never expected to visit again), and the South

Orkneys. On later AOs, we skipped those and went straight to South Georgia to save time. Near the South Orkneys lies the biggest iceberg graveyard in the world. Along most of the Antarctic coastline the ocean current runs from the east to the west. Transported icebergs keep hugging the coast, as the coriolis force tries to let them make a left turn (in the northern hemisphere coriolis makes thing turn to the right, including airflows, explaining the way depressions turn around). But further north, around the tip of the Peninsula, the current is west-east. There is a constant flow of icebergs leaving the Weddell Sea, which are picked up by the current and transported east. Coriolis makes them turn left and drift northeast all the way to South Georgia. Many of those strand at the South Orkneys, melting into the most fantastic, photogenic shapes. We sometimes make a zodiac cruise amongst them for the photographers. But we take care to only approach the small ones, and keep a distance from the big bergs. They look so serene and still, but they are inherently unstable. Everbody knows that food from the freezer thaws more quickly when you put it in water, which conducts heat much better than air. The same happens with icebergs, which melt faster under water than in the air. As a result, their point of gravity is constantly on the move, and when the underwater part becomes smaller they start to turn over. You often see former marks of the water surface running diagonally over the side. Sometimes this turning may go rapidly, and when the underwater part is raised into the air, it may fall apart, being carved by the sea. This process can accelerate, leading to the berg completely disintegrating in minutes. I have witnessed this phenomenon twice, the fist time when I joined Brazilian mountaineers on a glacier tour above their station in King George Island. Suddenly, our guide exclaimed "Vira, vira!" (it's turning, it's turning), pointing down into Admiralty Bay. And there, with a lot of noise, a large berg turned around and around, loosing pieces all the time, until after 15 minutes, nothing was left but a large field of floating brash ice. You don't want to be near such a thing in a zodiac. The second one I saw while sailing along the South Georgia coast, when we suddenly heard a loud bang, like a cannon shot. Then we saw a berg split in two, all the way from its peak to the sea, the two halves moving apart very slowly like a movie in slow motion. After a few minutes the two peaks reached sea level and the rolling around game began, both halves further desintegrating, but leaving much bigger chunks than the one in Admiralty Bay.

In South Georgia, we visit Grytviken, where the former Norwegian whaling station has been cleaned of asbestos. It has a nice museum and tourist shop, open all summer when there are about two ships a week visiting. We also toast on Shackleton's grave, and, having presented a lecture on him, I often had the honour of doing the ceremony. Everybody gets a shot of whisky, Shackleton's favourite drink, in a plastic cup. You are supposed to only take one sip, and then pour the rest over the grave, symbolising how much he liked it. I always wonder how a whole season of this affects the microfauna in the soil. We also make landings in various places to see penguins, albatrosses, and seals. The abundance of wildlife in South Georgia is bewildering and the scenery is unsurpassed. For me, South Georgia is by far the most beautiful place on earth I have ever seen (sorry Tristan).

On our way to Tristan da Cunha, almost a week sailing, with lots of seabirds around the ship all the time, we include a stop at Gough Island. We may not land there, the island being a strict reserve, but, when the weather is calm, we can make inshore zodiac cruises, sometimes coming close enough to see the endemic Gough Bunting, and we often see the flightless Gough Moorhens. The awe inspiring towering cliffs of Gough exude an unworldly, premordial atmosphere, and you would not be surprised to see flying dinosaurs any minute.

At Tristan we first visit The Settlement, see the museum, the pub, and the post office, and we try to make landings on the outer islands. Usually, the landing on Nightingale is fairly easy but Inaccessible is much more difficult. From the Nightingale landing place, where you are welcomed by the inquisitive thrushes, a very steep, and often slippery path leads up to the plateau. On our very first Odyssey, an elderly passenger slipped and fell, breaking his hip. Conrad Glass, in his book *Rockhopper Copper* (the only book about Tristan written by an islander) tells the story of how we had to get him in the zodiac. At Tristan the whole zodiac was craned onto the quay. Actually it was not an ordinary fracture. He had a steel replacement of the top of his thigh bone, fitting into the hip socket. The bone had split and the steel pin had come out sideways. This meant that he needed a full-scale operation, which had to wait until his return in England. The leg was stabilised in the Camogli Hospital on Tristan and packed in a plaster cast. We then sailed to Saint Helena as fast as possible, where he was pampered in the hospital for a few days. Finally we left him in Ascension where he took a flight home. We heard later that the operation had been successful and that he had fully recovered. When the Tristan guides on Nightingale heard about the hip replacement, I heard the most beautiful sentence of Tristan English: "They should 'ave toooold hus 'e 'ad n hartificial 'ip!"

After leaving Tristan, heading north for Saint Helena, we leave the Southern Ocean behind, and usually are in subtropical waters in subtropical weather under a blue sky within a day. We then also lose the albatrosses and petrels, with the exception of the Spectacled Petrels of Inaccessible. They stay with us for up to three days, hoping to pick up something edible, brought to the surface by the turbulence caused by the ship.

In Saint Helena we have an island tour, seing the houses where Napoleon lived, and his empty grave (the body has been exhumed and taken to France), and the birders always have a special tour to the breeding grounds of the endemic Saint Helena Plover, called Wirebird on account of its long, wiry legs. At sea, near the coast, dolphins are plentiful, and several times we have seen the giant Whale Shark, once even curiously circling the ship. At Ascension, the highlights are the nightly excursion to the egg laying turtles, and zodiac cruising at the foot of the whitewashed cliffs of Boatswain Bird Island, where the seabirds that have been wiped out on the main island, abound in their thousands, including Brown, Masked, and Red-footed Boobies, Red-billed and White-tailed Tropicbirds, Brown and Lesser Noddies, Fairy Terns with their translucent wings, and the endemic Ascension Frigate Bird. I will tell more about the various islands in separate chapters.

After Ascension we cross the Equator. And for some reason, I was elected to become King Neptune, a role I have played many times. The Russian sailors were used to having a rather rude way of celebrating. They would construct the 'tunnel', a wooden structure you could creep through, with rags hanging down, drenched in goo, more goo being added when the victim passed through. Then they would stamp the left cheek of your bottom with a trident rubber stamp, pulling your pants down, and slap you violently with other filthy pieces of rubber. We treated the passengers a bit more carefully, but nevertheless let them pass the tunnel, and we would stamp them, a bit less violently. Neptune was assisted by a team of evil looking devils, both Russians and Oceanwide staff members. The Russian devils would take care of crew members who needed to be baptised, the staff devils dragged the passengers to Neptunes throne a bit more gently. Albert Veldkamp had crossed the line many times as a whaler and he brought a certificate to prove it. But I declared him a fraud and sentenced him to triple treatment. After being treated the victims would be thrown into the pool to clean themselves. On the *Molchanov*, a makeshift pool would be erected on the stern deck, where we would have the ceremony, filled with sea water. On the *Plancius* we would use one of the zodiacs filled with water. At the end of the ceremony, when everybody was treated (except the cowards who never showed up), there was always a lot of goo left, and then the devils would grab Neptune, and pour all remaining filth over him and then throw him into the pool too. It is a special experience to slush around with the movements of the ship, together with other filthy bodies, in a luke-warm chocolate coloured soup with a nice cup of punch in your hand.

We once crossed the Equator in a total calm, not a whisper of wind. The sea was like a mirror, no waves, not even a ripple. We saw migrating Leach's Storm Petrels, from the Northern Hemisphere, flying around with perfect mirror images. As a great exception the captain allowed us to swim in the equator, lowering the gangway. Swimming in a 3000 m deep swimming pool, both sides being thousands of kilometres away, gives an eerie feeling. You almost expect a huge sea monster to come to the surface any minute.

After the Equator, we reach the Cape Verde Islands. In earlier Odysseys we would continue after that and end in either the Canary Islands or the Azores, but in later versions we ended the voyage in the Cape Verdes. This shortens the trip by almost a week, making it accessible for more people. Also, the most exciting things on this voyage happen in the first half, so it is not easy to keep everybody happy. Towards the end they start nagging about each other, the food, and about us. So stopping at the Cape Verdes is a good idea. Others leave earlier at Ascension, where you can take the rather expensive flight to Brize Norton near London. Cape Verde is an African country with African peculiarities, like a bus breaking down halfway through an excursion. Here we see how passengers who were perfect companions in the Antarctic may not be suitable for African tourism. I heard one American say:

"Look at it, it 's all garbage, all run down. Well, that happens when you give them independence."

And another one:

"Jesus! This is just a slave colony, left to breed!".

On one island excursion (with a bus breaking down) we offered the passengers a complimentary meal in a small restaurant in a quaint fishing village. People were not happy about the food, finding chewy pieces of pork with black hair stubble. In a co-ordinated action they demanded a five dollar refund. They pay thousands of dollars for the voyage, and now they want a five dollar refund; unbelievable. Cartoon of the day: a vertical cliff with a narrow road halfway, the bus breaking through the guarding rail, flying through the air into the bottomless ravine. A voice from the bus: "Do you think we can ask for a refund now?"

Generally speaking, the AO offers a great variety in islands, with very different climates, people, and flora and fauna. At sea, from Antarctica through the tropics, we see a huge variety of seabirds, and up to 16 different species of cetaceans (whales and dolphins). All my AOs have been very different. Every time I saw things I did not see on other voyages, and on every trip something goes wrong, but excitingly enough, you never know when or where. Landings may be impossible in bad weather anywhere, even at tropical Ascension, where heavy rollers, caused by storms thousands of miles away, can be very treacherous. When we introduced ourselves to the passengers, after leaving Ushuaia, I always liked to say that the first guaranteed landing on this voyage is the end point.

39. South Georgia

I am not the only one who thinks South Georgia is the most beautiful place on earth. Tim and Pauline Carr thought the same. In their book *Antarctic Oasis, Under the Spell of South Georgia*, they wrote:

"For twenty-five years our little sailing boat *Curlew* tracked across the oceans of the world and visited many countries, some with great individual charm. Always, though, we travelled on toward some elusive and ill-defined goal. More than a thousand miles east from Cape Horn and almost that distance northeast of the Antarctic continent, we reached the island of South Georgia and saw its awesome glacier-clad peaks rising steeply from the sea for most of its hundred-mile length. On closing with the coast the sounds and sights of tens of thousands of breeding birds and seals were overwhelming. We knew then that we had found the most exceptional place on earth. We fell under its spell and so here we decided to stay."

And that is exactly what they did. They had lived for decades, just the two of them, on their little wooden sailing boat without an engine, sailing the Mediterranean, the Carribean, the Atlantic Ocean from the Arctic to the Antarctic, around New Zealand and through the Pacific, but when they saw South Georgia they found their home. They arrived in 1992, and stayed there for 24 years, until old age forced them to leave in 2006 when they settled in New Zealand. In South Georgia Tim would do maintenance jobs for the government, while Pauline acted as curator of the Museum and worked in the tourist shop, having two or three ships visiting per week all summer long. They liked winters best when there were no visitors and they could roam all over the island on skis. We have seen the museum improve and expand over the years, its latest addition being a separate room with a replica of the *James Caird*, with a stuffed Shackleton at the helm. The *James Caird* was the boat used by Shackleton and a few mates to cross all the way from Elephant Island to South Georgia to get help for his marooned crew after the *Endurance* was lost in the Weddell Sea in 1915. The original *James Caird* is on display at Dulwich College in London.

The wildlife of South Georgia is overwhelming indeed. The shores are literally packed with penguins, Fur Seals and Elephant Seals, the tussock grass slopes dotted with various species of albatross. And, not visible in the daytime, myriads of assorted petrels burrow in the peaty soil. And all that against a background of mighty snowy peaks, reaching above 2,000 metres, and spectacular glaciers descending to sea level, producing small icebergs by calving. The big ones, always to be seen around the island, have drifted all the way from the large ice shelves of the Antarctic continent.

Elephant Seals are the filthiest animals I have ever seen. When they come ashore to moult, shedding the hairy outer skin layer, they lie in packs in their excrements surviving on stored fat, covered in filth, and their smell is unbelievable. During a visit to the whaling station of Stromness (now off limits for visitors because of asbestos danger), I found a group lying inside a derelict building, wallowing in their own filth, and the stench was so thick you could almost lean on it. And they are charmingly ugly. It is not hard to imagine that the old sealers gave them little respect. Lack of respect still occurs. I have been told that the funniest thing to do is the so-called elephant jump. Find an isolated sleeping Elephant Seal, and with one foot, jump onto the middle of his back, continuing with the other foot in a fluid movement to the other side. The startled animal will abruptly raise both the front and the back end, shaped like a huge banana, and if you step too far to the front or the back, you will be propelled into the air. I never tried it and I don't approve of it. Towards us, they are friendly and never attack like the seals do. If you come too close, which cannot always be avoided, they only open their big mouths, enveloping you in the worst breath you can imagine, and modestly shuffle a little backwards. They produce a sound best compared to loud human belching, with a flabbering proboscis (which gave them their name), always dripping with thick white snot. When the large bulls open their mouth they reveal powerful canines, with which they can seriously hurt each other when they fight over the rights to own a harem. They will raise themselves as high as possible, and use gravity to pound into each other with their massive bodies. Younger males, having no acces to the females, may satisfy their sexual needs with each other. I witnessed two males, lying in a sea-water bath, rubbing their bodies against each other, pink penises protruding in icy cold water. Unthinkable. A colleague of mine who spent many months studying them, managed to get a picture of an annoyed animal, smacking the pink thing of his neighbour away with his flipper, exactly at the moment it ejaculated a fountain into the air. A 'cum-shot', Elephant Seal pornography. The females and the young have lovely friendly faces with beautiful big dark eyes, the upturned corners of their mouths giving them a big smile, and they too have snotty noses.

The Fur Seals have made a remarkable come-back, after having been virtually exterminated by sealers in previous centuries, and over the almost twenty years I have visited the island I have seen them growing more numerous all the time. They are no longer confined to the beaches, but move further inland and uphill, as there is no room left on the beach. In places they displace penguin colonies, and pose a conservation dilemma. A similar situation arose in South Africa, where Fur Seals (another species) also came back in numbers, and now are a threat for the equally protected Cape Gannets. Young birds, just able to fly, alight on the sea, where they are grabbed and eaten by the seals. What to do if one red-list species threatens the future of another red-list species?

The Fur Seals on the landing places can be quite a nuisance. We tell our passengers to keep a distance of at least six metres from sea mammals, but the animals do not abide by the rules. Males can be agressive and charge at you, snorting loudly. If you flee, they may come after you at a surprising speed and try to bite your calves. Seal bites can be very nasty, leading to serious infections because seals never brush their teeth, and their mouths are very filthy, although their breath is not as bad as the Elephant Seals', smelling merely like fresh herring. It is best to make a firm stand, face the animal, stamp your feet and say "Go away!" They usually stop short, and you see them thinking "Oops," looking sideways, and with a last snort, more like a yelp, they turn around and indeed go away. But in general, you can just ignore them and their charges will stop short anyway. We never saw a seal bite. The females, nursing their young, totally ignore you when you walk past, completely absorbed by their maternal tasks. The little ones are extremely cute, with their baby faces and large dark eyes. They are very playful and when a gang of them come waddling towards you it is great fun to run away from them. They will all come after you as fast as they can, yelping loudly, not to attack, but just to play. Often a gang of youngsters comes splashing around the zodiacs during landing operations.

Of the seven species of penguin I have seen so far, my favourite is the Chinstrap Penguin, the species I studied at Elephant Island (not nesting on South Georgia, but sometimes visiting). They are the most endearing, with their eternal squabbles, and the focussed way they concentrate on their household tasks, always stealing the nicest pebbles from their neighbours' nests. I must confess I have been tempted to take them on the tip of my boot, making the flightless creatures fly, Moseley style, when I had once again sustained a painful bite, just below the kneecap, or my shins had been beaten black and blue by their powerful bony flippers. But I always realised I was the wicked intruder, weighing and measuring their innocent chicks, and they had every right to treat me that way. If a Skua made a low test flight over the colony, singling out an unguarded egg or chick, all the heads went up with loud shouts, causing a wave reminiscent of a football stadium. They respond much less violently towards the Sheathbills, white chickenlike birds, which move freely around in the rookery, feeding on penguin droppings and other disgusting things. They also scavenge, feasting on seal placentas, and they take penguin eggs, sneaking in when the owner is busy fighting his neighbour. Their filthy habits have caused them to evolve with naked faces, like vultures, so that they don't foul their feathers.

The true vultures of the Antarctic are the Giant Petrels. They have no bare vulture faces, and their heads may be covered in blood and grime up to the shoulders when they dig deep into the carcass of a dead Elephant Seal to get at the juiciest parts. They also take live penguin chicks, but with their long, straight, albatross-like wings, they cannot swiftly dive down to grab one, like the Skuas do. Instead, they walk in an ungainly fashion along the periphery of the colonies, dragging away unguarded chicks, proficiently turning them inside out.

The most beautiful and elegant of all penguins, no doubt, is the King Penguin, cousin of the larger Emperor Penguin of the deep Antarctic, famous from the film *March of the Penguins*. The Kings are more brightly coloured than the Emperors, and have a more slender build. They do not occur in the true Antarctic, but are confined to sub-antarctic islands. Like Emperors, they build no nests, but incubate their eggs on top of their feet, under a fold of skin, standing closely together. A rookery with a few hundred thousand King Penguins, packed together against a background of snow-clad mountains, offers one of the most impressive sights nature has to offer.

Of the four albatross species nesting in South Georgia, the Light-mantled Sooty Albatross is by far the most beautiful, dressed in amazingly subtle shades of grey. It is a close ally of the Peeo's of Tristan da Cunha and it performs the same spectacular aerial show, when two birds, in perfect synchrony, describe a huge horizontal figure eight in the air.

I can't resist repeating it once more: South Georgia is the most beautiful place I have ever seen.

Rats abounded around the old whaling stations but never reached the other parts of the island, as they did not cross the glaciers which come down from the snowy mountains, all the way to the sea. In the rat-infested areas burrowing petrels have been decimated, and the endemic South Georgia Pintail and South Georgia Pipit disappeared. Due to climate change, the glaciers are retreating at an alarming rate and might soon not reach the sea any more. Then there is the danger that rats may be able to cross in front of the ice and invade new areas, especially in winter when there is little flow of melt water.

The rats on South Georgia have been eradicated. It was a huge undertaking, involving ships and helicopters, dropping rat poison along precisely defined lines, using GPS. Between 2010 and 2015, more than 300 tons of poisonous bait were dropped at a total cost of 13 million pounds. In 2017 the island was declared rat free, and we saw Pipits and Pintails in places where we had never seen them before. Simultaneously, the Reindeer were exterminated. Reindeer were introduced by Norwegian whalers, to have something to shoot at in their spare time. Their grazing was detrimental to the indigenous vegetation, and their hooves trampled petrel burrows. Between 2013 and 2017 almost 7000 animals were shot, until none were left. After some of our visits to South Georgia, we would have a wonderful reindeer steak meal, on our way to Gough and Tristan da Cunha.

40. Saint Helena

I will not talk about Saint Helena's most famous inhabitant, Napoleon. Too many books have been written on him already, but none of them has the wonderful story about how his possessions were sold at the Bullock Auction Sale. Next famous is the oldest inhabitant of the island, Jonathan the giant tortoise. He was brought to the island in 1832 from the Aldabra atoll in the Indian Ocean, so in the next decade he will be approaching 200 years of age. He lives in the gardens of Plantation House, together with other, younger tortoises. Once we could not find him and we were afraid that he might have died. Then we heard a terrible noise from behind a bush, a low groaning "Aaaaaaargh". And there he was, mounting a beautiful young girl, perhaps not older than eighty. And, in the style of an old tortoise, he took it very slowly. Aaaaaaargh.

Saint Helena used to have a unique vegetation with many endemic species, of which preciously little is left. You can drive along the narrow winding roads for a whole day, enjoying the lush green landscape in the interior, with meadows, hedgerows, trees along the roadsides, and forests, without ever seeing one single original plant. Everything you see is imported, replacing the original vegetation. Only a few small patches remain, mainly along the humid, foggy central ridge at about 800 metres altitude. The endemic plants have confusing local names. There is an Ebony, which is not related to the African Ebony. There is a Redwood which is not related to the Redwoods of California, There is a Lobelia which is not a Lobelia al all, and a Rosemary which has nothing to do with our Rosemary. There are Cabbage Trees which are not cabbage of course, with funny names like the He Cabbage, the She Cabbage, and the Black Cabbage Tree. Strangely enough, these trees are in the daisy family, which we mostly know as herbs, like daisies and dandelions, the largest being the sunflowers. The Cabbage

Trees have inconspicuous inflorescenses, but the Gumwoods have beautiful daisy-like white flowers. There is also a False Gumwood and a Bastard Gumwood, more strange names.

There are two main categories of endemic species. There are the ones that colonised islands, leaving their ancestral stock on the mainland unchanged, while on the islands they evolved into something different. Birds became flightless, for instance. In the other type, the forms which came to the islands remained unchanged, while their ancestral stock on the mainland became extinct, being replaced by new, more competitive species. The latter we call 'old relic endemics'. The daisy trees of Saint Helena are thought to be of this kind.

A strange thing happened with the Ebony and the Redwood. Both were cut down in the past, and Redwood beams can still be found in the ceilings of old houses in Jamestown. In addition, they succumbed to grazing, against which they had no defence. The last full grown Redwood died some years ago and now there is only a handful of low shrubs left, as if they know that growing up is dangerous. You will be chopped down if you do. Also, the remaining shrubs would not multiply anymore, pollination failing, and cuttings would refuse to grow up. So the species was on its way to extinction. Finally, with hand pollinating, seeds could be produced, and methods were developed to ensure cuttings survived. So now Redwoods are cultivated in numbers and replanted, but they still don't grow into real trees. Ebony, prized for its hard black wood, used for ornamental purposes, was believed to be totally extinct for more than a hundred years, but two bushes were rediscovered in 1980 on a cliff side, by Quentin Cronk and George Benjamin, the leading botanist of the island. Benjamin's brother was lowered on a rope to take cuttings, from which the species has been cultivated since, side by side with the redwoods. Then a remarkable thing happened. Ebony and Redwood are closely allied species, with beautiful bell-shaped white or pink flowers, and they found each other, forming a bastard. In contrast to their slow growing parents, the bastard grew vigorously. It was named Rebony and is now an ornamental shrub to be seen in many gardens on the island.

The Gumwood used to be a major element in the original forests, and of all the rare endemic trees of Saint Helena is still the most common. The Bastard Gumwood was thought to be extinct, but like the Ebony it was rediscovered. The last known living tree grew at Longwood (Napoleon's house) but was blown down in a gale. It was then considered extinct, until Stedson Stroud found a single tree in 1982. He managed to cultivate it from cuttings. The False Fumwood was down to a handfull of individuals, but propagation from seeds and cuttings is now very successful. And, as happened between Ebony and Redwood, False Gumwood and Bastard Gumwood bastardised. A bastard between false and bastard. How false can a bastard be?

The invertebrate fauna of Saint Helena contained a remarkable number of endemic species, especially amongst the spiders and beetles, most of them extinct. The most famous one was the Giant Earwig, which grew to a length of five centimetres, and was last seen alive in the 1950s. Of the many land snails, only a few survive, including the pretty Blushing Snail, with a translucent reddish brown shell, which gave it its name. George Benjamin used to show us a small patch of Black Cabbage Tree woodland, with a nice undergrowth consisting of endemic ferns and herbs. He would turn their leaves, to show us the Blushing Snails, and the most curious of all Saint Helena creatures, the tiny Spiky Yellow Woodlouse. Saint Helena has an arboretum, where most of the endemic trees have been planted, named after George Benjamin. He died in 2012 at the age of 76.

There were also endemic birds, including two kinds of flightless rail, one of which was classified as an *Atlantisia* species, in the same genus as the Inaccessible Island

Flightless Rail. The only endemic bird surviving is the Saint Helena Plover, locally known as Wirebird. Wirebirds developed into true meadow birds, depending on grazing. As I am a specialist in Dutch meadow birds, I dubbed it 'the loneliest meadow bird on earth'. It is the 'national bird' of Saint Helena, depicted on the archway leading from the landing pier into Jamestown. All the birds you see twittering and fluttering in town are imported species from around the world, with the exception of the white Fairy Terns with their translucent wings, which nest high in the trees of the Castle gardens and on narrow ledges on the church. Red-billed Tropicbirds nest in the vertical cliffs either side of town, where cats and rats cannot reach them. Other seabirds are confined to offshore stacks, being wiped out on the main island. Bone deposits show that large colonies of petrels and shearwaters used to live there, including the extinct Saint Helena Petrel. An unidentified petrel has been seen twice and even photographed in 1988. Several people have been to the ridge where it had been seen to listen for calls during the night, but failed to make more observations. But you never know. The related Cahow of Bermuda, or Bermuda Petrel, was thought to be extinct for more than three hundred years, until it was rediscovered in 1951. Zino's Petrel, or Freira, was thought to be extinct and was rediscovered in 1969 in the highest ridges of the mountains of Madeira.

Saint Helena has the reputation of being impossible to escape from, if you are a prisoner (a nice reason to put Napoleon there). Yet, it has happened. In the museum there is a newspaper article on display telling the story. A Dutch captain was caught with a large quantity of marihuana on board, and jailed. Apparently he was not carefully guarded, because he managed to poduce a key in the prison workshop, which enabled him to leave the prison at night, when nobody was watching. He roamed about freely in Jamestown, unnoticed, always returning to his cell in time. He even made a friend who was willing to help him with an escape plan. He first thought of stealing a yacht, but decided it would be too risky. So he let his friend build him a small boat. On a dark night he sailed away, with food provisions and water. When his disappearance was discovered the following day, nobody thought he could have left the island, as no boats were missing, so all searches were limited to the island itself. Meanwhile, he drifted west with the trade winds and the currents, but is was not an easy voyage. Not being used to sunshine, he was terribly burnt, and his progress was slower that he had anticipated, so he ran short of food and water. He finally washed up on the Brazilian coast, more dead than alive, where he destroyed his boat. He found a woman who pitied him and took him in. After he recovered, she even gave him money to travel to Brasilia, the capital, where he went to the Dutch embassy. Possessing marihuana is less of a crime in Holland than in Britain so it was decided he had served enough time in Saint Helena. He was sent to the Netherlands as a free man.

Saint Helena has its own Robinson Crusoe story, one of my favourite 'gruesome stories'. Fernão Lopes was one of the 1300 soldiers sailing with the fleet of Tristão da Cunha in 1506. The Portuguese occupied the coast of India, with the exception of the fortified Muslim enclave of Goa. They finally conquered it, and Lopes was one of the soldiers guarding it. But Rasul Khan (Roçalcão in Portuguese) took it back, giving the Portuguese occupants the choice between being killed or convert to Islam and become his soldiers (like Thomas Swain became a French soldier). Albuquerque was furious and tried to regain control over Goa. It ended up in a prolonged siege with great losses on both sides. Eventually, Goa was handed back to the Portuguese in a peaceful way.

In the peace negotiations it was agreed that the renegade converts would not be killed. So the lives of Lopes and his friends were spared, but don't ask how.

First, black henchmen plucked out all their hairs, not very gently, including all the hairs on their bodies, eyebrows and eyelashes. Raw and bleeding, they were covered in Pigs' manure until their faces were no longer recognisable, and anybody who wished to urinate on them was invited to do so. The next day their noses and ears were cut off. The wounds were not treated and they were thrown in pig manure again. On the third day their right hand was chopped off, and the thumb of the left hand. These wounds were dressed to make sure they would not bleed to death. The next day they were turned loose. Many quickly died, often with terrible infections. Lopes survived as a beggar living on garbage. In 1515 he found a captain who pitied him and was willing to take him to Portugal, where his wife was still living. But when they reached Saint Helena he was stricken by fear that he would not be welcome at home, and that his treatment in Portugal would not be good at all, so he fled ashore and hid himself in the forest. The crew could not find him and he was left on the island. The crew left him some provisions and clothes on the beach. Saint Helena had not been colonised yet, and Lopes became the first permanent inhabitant. He may have had a black slave as companion for a while, and it has been suggested that this man was the role model for Friday in Daniel Defoe's story of the real Robinson Crusoe. Over the years, a situation developed where Lopes would deposit fresh fruits and vegetables on the beach for visiting ships, who would leave him clothes and other things in return. He always stayed hidden in the forest during these visits. He probably lived in solitude the first ten years, and then found company in a young man of Indian descent. This man betrayed his secret dwelling to the captain of a passing ship, who tried to persuade Lopes to join him to Portugal. Lopes had become quite a legend, and the King had expressed the wish to see him and thank him for what he did for ships visiting. Lopes refused at first, but eventually gave in. The King received him, and arranged a visit to the Pope, who gave him absolution for his sins.

Lopes was not happy in Portugal. He developed a fear of crowds and noise, and in the end he had only one wish: to be returned to his peaceful island. In 1530 or 1531 he was taken back to Saint Helena, where he lived in the forest for another fifteen years. His body has never been found.

I have been to the French Island La Réunion, in the tropical Indian Ocean, one of the Mascarenes, east of Madagascar, at about the same latitude as Saint Helena. The Mascarenes (Réunion, Mauritius, and Rodriguez) used to have more endemic creatures than the Galapagos Islands (including the Dodo from Mauritius and the Solitaire from Rodriguez), but most of these have gone extinct, following the colonisation by Man. Réunion still has a number of endemic bird species, including the mysterious Barau's Petrel, which was only described as recently 1963, nobody knowing where they were nesting. Various expeditions failed to find their nests. Only very recently, in 1995, has it been established that their colonies are in the elfin forest above 2400 m altitude, above the thick clouds, which make Réunion one of the rainiest places on earth. They are fairly easy to see, when in the afternoon they fly inland, to reach their colonies high up the mountains. Most related petrel species are nocturnal, but Barau's Petrels return to their nests in the daytime, making use of thermal updrafts to gain altitude. Réunion also has a few patches of indigenous trees left, most of the forests having been cleared in the past. There I was surprised to see trees that looked very much like the She Cabbage of Saint Helena. I never heard of any kinship between daisy tree species of the two islands, having the whole mass of Africa between them, but I think it is quite likely that the daisy trees on Réunion, like those on St Helena, are endemics of the relic kind, remnants of a much wider distribution in Africa in the distant past. Similarly, Réunion hosts relatives of the Saint Helena Redwoods and Ebony.

In the Indian Ocean, I would also have liked to visit

Amsterdam Island, the mirror image of Tristan, at the same latitude, with the same trees, tussock, penguins, albatrosses, and rocklobsters. Amsterdam is part of the *Terres australes et antarctiques françaises* (TAAF), together with nearby Saint Paul, the Crozet Islands further south, and Kerguelen still further south. Like Amsterdam corresponds with Tristan, the Crozets correspond with the South African islands Marion and Prince Edward, with large populations of albatrosses and penguins. Ile de la Possession, in the Crozets, hosts the largest King Penguin rookery in the world, containing about half the world population. Kerguelen corresponds with South Georgia and is much larger. The main island, Grande Terre, covers 6675 square km, with glaciated mountains, and is deeply indented by fjords. It is surrounded by about 300 smaller islands. All islands of the TAAF are uninhabited, but have scientific bases, which are served four times per year by the ship *Marion Dufresne*, based in Réunion. Apart from the technicians and scientists going to the various stations, the ship has place for about twelve passengers. I have met several people who have made the voyage, often French passengers we had on board, and they especially praised the food and wine on the ship. At least four times I have tried to obtain a berth, but failed to do so. My last attempt was in 2020, but they would not take any passengers for the whole year, because of the Covid-19 pandemic. And then, in the same year, I passed the maximum age they accept on board. Amsterdam Island has to remain a dream.

Saint Helena has its own Moorhen, corresponding with the South African subspecies of the Common moorhen. It lives in densely vegetated valley bottoms and can still fly. But give it a million years, and it may develop into a ferocious flightless Island cock...

41. Ascension

Like Tristan da Cunha, Ascension Island is part of the UKOT Saint Helena, with an administrator responding to the Governor in St Helena. Tristan, St Helena and Ascension not only share a governor, but also a flightless rail in the genus *Atlantisia*. The Ascension Rail, *Atlantisia elpenor*, is a close relative of the Saint Helena Rail and the Inaccessible Rail. And like the Saint Helena Rail, it is extinct. It has been seen alive by a human being only once, way back, in 1656. Traveller Peter Mundy describes it as follows:

"...alsoe halfe a dozen of a strange kind of fowle, much bigger than our sterlings or stares: collour gray or dappled, white and black feathers intermixed, eies like rubies, wings very imperfitt, such as wherewith they cannott raise themselves from the ground. They were taken running, in which they are exceeding swift, helping themselves a little with their wings (as it is said of the estridge), shortt billed, cloven footed, thatt can neither fly nor swymme. It was more than ordinary dainety meatt, relishing like a roasting pigge."

Mundy even adds a drawing of the bird, that in my view does not really look like the Inaccessible Rail, but more like our Corncrake. He then starts to philosophise about the origin of such strange birds, like the Dodo of Mauritius, which also can neither fly nor swymme. How did they get to the islands? Were they created by God on the spot, or had they come from elsewhere and changed in their appearances and capacities? In fact, Peter Mundy postulated the possibility of evolution, two centuries ahead of Darwin.

Nothing was heard again of the rails on Ascension until three centuries later when the British Ornithologists Union (BOU) organised an expedition to Ascension to celebrate their centennial. In September 1958 they found bone deposits in a fumarole. Fumaroles are a special kind of volcanic crater that do not produce lava, only sulphurous fumes. Birds coming too close to the edge may get nauseated and fall down into the pit. The ornithologists found many bones of seabirds, but also a skull that unmistakably belonged to a rail.

In 1970 and 1971 Storr Olson from the Smithsonian Institution (USA) examined the fumaroles of Ascension more closely and found bones of more than 50 rails. He formally described the species as *Atlantisia elpenor*.

Ascension is a strange island. Officially it is uninhabited, as the thousand people who live there are contract workers or their families. When contracts end people have to leave. Many of those are from Saint Helena. Ascension was, like Tristan da Cunha, occupied by a garrison when Napoleon was imprisoned on Saint Helena, but on Ascension the military stayed forever. The status of the island was not a rock, but a ship, called HMS *Ascension*, sloop of war. Ascension has a military airport, shared by the British Royal Airforce (RAF) and the US Air Force, who have a big say in what is happening there. It has one of the longest runways in the world, and in case of emergency it can accomodate the space shuttle landing. At the beginning of the 21st century an attempt was made to change the peculiar military status of the island and to allow people to own private property, but soon that decison was reverted. I don't know what exactly happened, but I suspect that the Americans (the Bush Administration) did not want it.

Ascension used to host large colonies of seabirds, but they were exterminated when cats were introduced to combat the rats in 1815. Instead of killing the rats, the cats took the seabirds. Guano deposits are still the silent witnesses of the once abundant birds. The only seabird surviving in large numbers was the Sooty Tern, locally known as Wideawake. Their strategy was to be totally absent for three months between their breeding seasons, thus limiting the number of cats living on them. Then they would arrive in millions, and although the cats would kill them by the thousands, this would not affect their numbers too much, simply because there were too many of them. This phenomenon is called swamping. But now the cats are gone. In 2004 and 2005 all cats were killed by trapping, shooting and poisoning. This was a great achievement and the seabirds which had abandoned the island, only surviving on offshore islets like Boatswain Bird Island, started to nest on the main island again. In 2006 one of our Odyssey passengers was 'Dave the Cat Killer', who had been responsible for the eradication, sponsored by the Royal Society for the Protection of Birds (RSPB). Dave was not very popular on the island, as about one third of the so much loved domestic cats were also killed accidentally. Cats and dogs on the island have to be chipped and neutered nowadays.

Ascension is a desert island, mostly consisting of bare volcanic rock and cinder cones. In the centre a peak rises to above 800 metres, called Green Mountain, as the peak catches clouds and humidity. When the soldiers occupied the island they established a farm near the summit where vegetables were grown for the regiment, and a water catchment basin was built, piping the water down to the settlement of Georgetown, as there were no other water sources anywhere else on the island. Attempts have been made to make the island greener. In 1845 many plants were imported from Buenos Aires, followed by 700 packets of seeds from Kew Gardens in London. In 1858 228 new species were introduced from South Africa, and over the following twenty years, 5000 trees were planted. As a result of all these efforts, very little is left of the original vegetation, with its endemic elements. In the Farm vegetables are no longer grown, all foods being imported. Instead, Stedson Stroud from Saint Helena, the man who rediscovered the Bastard Gumwood, has his greenhouses there, where he cultivates endemic fern species which he replants in nature.

Ascension has become noticeably greener in the last decades, especially after the introduction of Mexican Thorn, or Mesquite, a thorny North American desert tree, which has spread all over the island. It is related to, and superficially resembles, the equally thorny Acacia species of the African savannas. Being drought resistant and fast growing, Mesquite has been widely planted in forestry projects in Sahelian Africa, for instance in Senegal, Mali, and the Cape Verde Islands. But on Ascension is is not welcome. On board, I used to show a lengthy video about the island, including an interview with the Administrator, who complained about tyre punctures. He had no idea how the awful tree had come to Ascension, and was investigating possibilities to get rid of it. A snickering sound came from the back of my audience. There was this elderly British man, a retired landscape architect. He had done interesting jobs, like re-landscaping the area around Mount Pleasant Airport in the Falklands after the war with Argentina. He had planted them on Ascension along a steep roadside to stabilise creeping soil.

The barren red and black rocky island is fringed by amazing golden beaches. These have been produced by predatory fish, which chew up shells and corals, digest the contents, and excrete the rest. Thus, over a million years, their excrements accumulated to form the golden beaches. And amazingly enough these beaches have been found by sea turtles, which graze the seagrass beds of Brazil. They swim two thousand kilometres to these beaches to lay their eggs. The British developed a profitable business, producing turtle soup, until over-exploitation led to the collapse of the industry, just like with sealing and whaling. When the Americans arrived in the Second World War, remaining forever, they fortunately did not like turtle meat. The numbers have increased, they are well protected nowadays, and watching egg laying turtles has become a tourist attraction.

Like Saint Helena, Ascension had its own Robinson Crusoe story, another one of my gruesome stories. In

January 1726, British visitors found an abandoned tent, and a diary, written in Dutch, starting May 5th 1725 and ending half a year later. The journal was translated and published as *An Authentick Relation*. The original Dutch manuscript has been lost.The author, a Dutch sailor, was left on the island after he had been caught in the act of sodomy with a fellow crew member. He survived on a diet of birds and turtles, mainly drinking their blood, as there was no water. He finally succumbed to dehydration. His diary did not name people or a ship, so discovering the historic details would not be easy. In 1978 a so-called historic version was published by the American author Peter Agnos, called *The Queer Dutchman.True account of a sailor castaway on a desert island for 'unnatural act' and left to God's mercy*. Agnos gave the man a name, and added scientific footnotes and references to archives. It all appeared to be fantasy. He made it all up but he revived the legend. Dutch historian Michel Koolbergen found the real truth in the East India archives in The Hague, but he died in 2002 before he could publish his findings. In 2006 his book was finally edited and published by Alex Ritsema, *A Dutch castaway on Ascension Island in 1725*. Ritsema called this his contribution to 'Pink literature'.

Leendert Hasenbosch worked as a clerk on board the *Prattenburg*, sailing from Batavia to Holland in 1725. Between Saint Helena and Ascension he was caught having sex with a friend. On the 5th of May he was put ashore on Ascension, which was then still uninhabited and claimed by no country. He was left with a tent, a barrel with drinking water, two buckets, and a quantity of onions, beans and rice. When he saw the water level in his barrel sinking, he searched for water sources on the island. He followed herds of wild goats, hoping they would guide him to a well, but to no avail. He ate birds and turtles, but thirst became his gravest problem. He bled turtles in his bucket and drank their blood, but it made him sick. He had to throw up and suffered from diarrhoea, which aggravated dehydration. He opened the turtles' bladders, finding their urine more palatable to drink than blood. He experimented with mixtures of blood, turtle urine, and his own urine, but in the end he had to give up, and died around the 14th of October. The last entries in his diary only mention "the same, today". Three months later, the English sailors found his tent and diary. His body was never found.

42. Quincentenary

In 2006 we celebrated that 500 years ago Tristão da Cunha discovered the islands. For the occasion, the RMS *Saint Helena* was chartered for the Tristan da Cunha Association to make a special trip to Tristan da Cunha, in January, the height of summer. It would take about three weeks: almost one week sailing from Cape Town to Tristan, then we would spend a whole week at Tristan, and again almost a week sailing back. It was announced in the Newsletter, and I made sure I was on the passenger list (this time paying for myself).

As with the 'Old *Plancius'* and the 'New *Plancius'*, there was an old RMS *Saint Helena* and a new one. The old one sailed from 1978 until 1990, when she was replaced by a new ship with the same name. She sailed bi-monthly from England to Cape Town, making stops in the Canary Islands and Saint Helena. From Saint Helena, it would run a shuttle back and forth to Ascension Island. Once a year it would make a detour via Tristan da Cunha. However, when ownership changed and the southbound voyage ended in Walvisbaai instead of Cape Town, the detour via Tristan was deemed too long, and the annual visits ceased, leaving the visit of the SA *Agulhas*, once a year, as the only regular service, supplemented by the much more irregular sailings of the fishing vessels.

In 2005 the construction of an airport on Saint Helena was approved. RMS *Saint Helena* was scheduled to make her last sailing in 2010, when the airport was supposed to be ready, but due to endless problems around the construction, this was postponed year by year, until finally she made her real last sailing in 2018. A regular flight service between Saint Helena and Ascension might look the most logical thing to do, as so many Saints have a job on Ascension. But instead, the island has an air connection with Johannesburg (the USA blocking things again?). If you want to travel from Ascension to Saint Helena, you first fly from Ascension to London, then from London to Johannesburg, and finally, from Johannesburg to Saint Helena. The RMS *Saint Helena* was almost never mentioned by her full name. She would usually just be called 'The RMS'.

It was a memorable voyage and we had the most interesting passengers, such as the Governor of Saint Helena who came especially for the celebration. Then there were all the living members of the Gough Island Scientific Survey, half a century ago, including expedition leader Martin Holdgate, and, of course, ornithologist and eco-criminal Michael Swales, who shouldn't 'ave dunnit. He now speaks of his successful re-introduction of the Moorhens, rightfully considering the two forms of Tristan and Gough the same species. Holdgate climbed high, after Gough, becoming president of the IUCN (International Union for Conservation of Nature, headquarters in Gland, near Geneva, Switzerland). Swales, apart from his interests in ornithology and conservation, developed a deep affection for the people of Tristan, almost to the level of becoming a 'secondary godfather' after Allan Crawford. He arranged that young people from Tristan could spend time in the UK, studying at Denstone College. When I visited Swales in 1993 it was still a boys' school, but in the meantime it had evolved into a mixed gender school. Pupils from Tristan are often girls. Officially it was called an exchange programme, but I don't think students from Denstone College ever went to Tristan for their education.

On board, the expedition members, all in their seventies now, were affectionately known as the 'Goughalites'. The only Goughalite missing was Nicel Wace, the botanist. He was a professor in Australia and had just recently died. The others brought his ashes, with the intention of depositing them on Gough Island, but they did not get permission to land there. So instead, they left half of the ashes on Tristan, on one of Nigel's favourite spots, and sprinkled the rest in the sea near the Glenn, where they had their base at Gough Island, when we circumnavigated that island on our return voyage. And we all observed a minute's silence.

There were all sorts of passengers. Of course there was a fair number of stamp collectors. One of them bought the entire stock of covers for sale in the ship's shop the minute it opened on the first day of sailing. Others were furious. This man and his wife were called 'The Grabbers' for the rest of the voyage.

Then there was a London underground tube driver. As a little boy, a globe balloon was hanging above his bed and when he looked up, lying down, he saw the underside of the Atlantic, marked by this little magic dot. Hooked for life. And a young woman who had been violently robbed in her home, and developed a severe case of agoraphobia after that. It grew worse, until she didn't dare to leave the house, or even open the curtains anymore, living in the dark. She sought psychiatric help, and as a remedy to be amongst people again, she was advised to join a photo club, where she could meet people in darkness in the darkroom. It worked. She found her partner there, and now, emerging from her self-imposed isolation, as part of her healing process, she wanted to see the most isolated village in the world, and write a book about her experiences, entitled *From isolation to isolation*.

There were also ex-administrators and people who had worked in other professions on Tristan in the past. And there was Cathy Snyder, daughter of Peter Munch, who joined the Norwegian expedition almost 70 years ago. She now lived in the USA, and worked for the FBI as an IT-specialist, connecting all sorts of data bases. The first real spy I ever met. She had her father's diaries, which she intended to publish. But because they contained sensitive information about other expedition members, she did not want to do that whilst they were still alive. So she had to wait a little longer, because Allan Crawford was still alive (he died one year later). *Glimpsing Utopia* (with a wink to the title of her father's book *Crisis in Utopia*) was finally published in 2008.

Also on board was Sarah Sanders of the RSPB (Royal Society for the Protection of Birds). She travelled to Tristan to discuss various conservation issues, mainly concerning the influence of rats and mice. She presented a lecture on the house mice of Gough, which had recently become a serious environmental threat. They had always been seen as bothersome (at the weather station) but innocent. But South African scientists discovered that they developed the habit of attacking albatross chicks at night, nibbling at their abdomens, until they started to bleed. The mice would suck on the wounds, which would get infected in due course, so the mice could feast on blood and pus. Sarah showed horrible pictures, taken with infrared cameras, where mice were even hanging on the eyes of the albatross chicks to lick their tears. And the poor birds would not do anything to defend themselves. After weeks of suffering they would succumb and die. Sarah generously gave me permission to copy her lecture for future use on my Odysseys. Thus, I added another gruesome story to my collection.

In order to safeguard the future of the Tristan Albatrosses on Gough, an eradication programme would be necessary. Eradication of mice and rats on islands has become a New Zealand specialty, with various successful examples. A specific bait, only eaten by rodents, has to be distributed all over the island, using helicopters, following a narrow grid using GPS navigation. In inaccessible places, like caves, additional bait has to be deposited by hand. It is a complicated logistic operation, and the eradication of the mice on Gough was initially estimated to cost around three million pounds, but I am afraid the price has gone up since. Fund raising had already started but the project was postponed when the rats on South Georgia were given priority (see chapter 39).

Years later, I heard that exactly the same thing happened on Marion Island in the southern Indian Ocean. Like on Gough, the mice learned to feast on albatross chicks during the night. It was seen for the first time in

2009. While the mice on Gough mostly attack the abdomen, the mice of Marion climb on top of the heads of young albatrosses and gnaw away the skin, completely scalping them. Footage on the internet is even more gruesome than that I have seen from Gough. The Megamice of Gough have received more attention in the international press, but the situation on Marion is just as serious. The South African government and Birdlife South Africa are preparing for an eradication scheme similar to the ones on Gough and South Georgia.

There were also other Dutch people on board. Sandra Kornet and her husband Albert, who were at Tristan in 2001 to study the past of Peter Green, could not resist the temptation of visiting the island again.

We even had a true episode of cruise-ship soap on board. Like Alexander Greig with the *Blenden Hall*, I will not use real names. Let's call them Mrs She, widow, and Mr He, widower. They had a relationship but she kicked him out when she found out that his wife was not dead. Then his wife died, and he came back to her on his knees. She let him in (the stupid cow, someone else on board remarked), and now they were celebrating their reunion on this trip. But they started to fight. He especially despised her drinking behaviour. So they split up. As all the cabins were occupied, he was given the hospital bed to sleep in. She sought the company of her drinking buddies (like me), explaining what an asshole he was, while he told the other passengers what a bitch she was.

From Cape Town to Tristan we had the most wonderful sunny subtropical weather, all the way, through calm seas, Totally different from my stormy passage in 1993. We even had a Red-billed Tropicbird following the ship for a while, a true tropical species. So we were outside all the time, on the sun deck and in the swimming pool. I shared a cabin with two nice, elderly Irishmen. In the evenings after dinner, we were entertained by staff members with all sorts of games, unknown on Atlantic Odysseys. I especially remember Sharon and Aaron, Purser and Steward, both Saints, always in impeccable white uniforms. I met them both on later visits to Saint Helena. Sharon once joined us on a dolphin trip with the *Gannet III*, as she was a friend of Skipper Graham. She looked completely different in faded jeans. I met Aaron when I took a taxi to Sandy Bay, and the driver turned out to be his grandfather.

On Tristan we were given the opportunity to spend a few nights ashore. Half the houses of the island were converted into B&Bs, a totally new phenomenon on Tristan. It appeared that most people had pre-arranged their stay onshore. I had overlooked the information regarding this possibility and I had not. Fortunately, my old friend Joyce Hagan managed to arrange a bed for me, so I stayed with her, sharing with three Goughalites, including Martin Holdgate. Professor Edi Gittenberger from Leiden University (who finally adopted my Moorhen case as a DNA study), had asked me to try to collect endemic *Balea* snails for him. There are several species living on the islands of the archipelago, including Gough, so Holdgate, the zoologist of the expedition, knew them very well. There is even one named after him, *Balea holdgatei*. Another one has been named after Michael Swales, *Balea swalesi*. Holdgate was quite willing to help me, so we spent a whole afternoon together, looking for them. We failed to find them. We saw many snails in different forms, even *Balea* lookalikes, but they were all alien, introduced species.

Back on board after my stay ashore, I met my two cabin mates. One of them said: "Guess what!", and when I looked at them questioningly, the other said: "They are back together!" And indeed, at dinner Mrs She and Mr He were sitting at one table, apart from everybody else. She had alienated half the passengers, he the other half, leaving nobody for them to speak to on the return voyage.

Apart from the horrible story about the mice of Gough Island, another gruesome story came to the surface on this voyage. A philatelist had found out that in 1958 the USA had tested atomic bombs in the atmosphere above the South Atlantic, not far from Tristan. The news circulated among stamp collectors, and with us, reached Tristan da Cunha. Everybody was shocked. The Goughalites, realising that this happened only shortly after their expedition, were suddenly aware of the possibilty that the fact that they had been pushed to go to Gough instead of the Arctic, and that the Royal Navy had been so willing to give them logistic support, might have something to do with this. Perhaps the whole idea of this expedition was to collect baseline information on the life on Gough Island in case something went terribly wrong. It was also said that British authorities had denied knowledge of the explosions. I don't know if this is true, and I did not check it. Furthermore, by coincidence (?) the fishing vessels were called back to South Africa for maintenance during the experiments. It may all seem like a big conspircy theory, the only certain fact being that the explosions took place. The Goughalites were appalled by the idea that perhaps they may have been used for this purpose. Another gruesome story for my collection. Combining the two, we made the joke that the mice of Gough had not gradually evolved to become these terrible Megamice, but that their change was due to a mutation caused by radio-active radiation. We have called them the Giant Mutant Mice of Gough ever since.

All week we had nice weather on Tristan, and we also went over to Nightingale, not with the RMS, but with the fishing vessel *Edinburgh*. We had good views of the thrushes and the buntings, including the rare Wilkins' Bunting in the *Phylica* forest near the top. In the late afternoon we saw the Greater Shearwaters coming back to the island in their hundreds of thousands, a phenomenon I had not seen before. With our Odysseys we are always too late in the season. From a distance it looked like a swarm of mosquitos.

In spite of nice weather, there were always clouds below the edge of the Base. Except on the last day. Suddenly there was a totally blue sky, the top of the cliffs in sharp contrast delineated against it, like on the day I went to the peak in 1993. And, again, I could not resist the temptation. I wandered away from the village, uphill, pretending I was just loitering, looking down all the time to see if anybody was watching. Nobody ever looked up, so I continued, eventually disappearing from view. I soon found the Hottentot Gulch trail, which I followed all the way up to the Base, where I had some wonderful encounters with Mollies. In the gulch I met a family of Tristan Thrushes, and the view down was magnificent. The Settlement seen from above is a classic picture, often seen. I could not stay long, because we had to board that afternoon. In the harbour Jeff came to me with a grave face. He said that I (of course) had been seen, and my behaviour was utterly irresponsible. I said he was right, and then we shook hands, saying "take care" to each other.

Three months later I was back on my next Odyssey. This time all places on board had been sold, so when we had to take eight island guides on board to go to Nightingale, eight people from crew or passengers had to stay on Tristan, to comply with international safety standards. Some crew members volunteered, including our Philipino household staff, who were not really interested in birds. One passenger, a German mountaineer, would rather stay on Tristan and climb to the Base, if he could find a guide. I also volunteered, as I had been to Nightingale just three months ago, and weather conditions this time were much less favourable. We discussed the guide for the German with Connie Glass, and of course there was nobody available. Then Conrad said, to my great surprise, that I was allowed to take him to the Base, providing we kept radio contact. "You know the way," Conrad said simply. That was a great moment in my Tristan career.

Finally acknowledged! The German and I went up Hottentot Gulch at a relaxed pace, and spent several hours on the Base, where I had the best encounter with an Island cock I ever had, giving me great photo opportunities. We saw several Yellow-nosed Albatross chicks, sitting on their nests, and they even had rings on their legs. So the Molly Plot People had been up here!

The *Molchanov* returned late, and when we saw her approaching Tristan we noted that she did not come from Nightingale, but from Inaccessible! It turned out that the sea and wind were all wrong at Nightingale, and that a landing was not possible. So they turned to Inaccessible instead, and in an almost suicidal attempt managed to get ashore. The eight island guides were all up to their waists in the water to keep the zodiacs under control at the landing place. But the birders saw the rail, and I was not there! That evening, the atmosphere on board was unbelievable. All tension gone. They all made fun of me for missing the bird, but I thought my time would come, and indeed, I had my landing on Inaccessible the year after and I saw the bird. We had a terminal cancer patient on board. He had quit his job, sold all his possessions, and started to travel to see the birds of the world. He was giving away beers left and right, and with tears in his eyes, he said that this was the most beautiful day of his life.

43. Operation Argus

In the 1950s the Cold War between the West and the East was in full swing. In contrast with this, both Russia and the USA participated peacefully in the International Geophysic Year (IGY) 1957, together with 65 other nations worldwide. The idea was to gather scientific data on the properties of the earth, including Antarctica and outer space, in a co-operative non-military atmosphere. But of course the great powers had their secret second agendas. During the IGY, the Russians launched the first satellite in history, Sputnik 1, on the 4th of October. The USA panicked. They were not ready yet with an answer and feared they had lost the race for space. To make matters worse, the Russians launched Sputnik 2 less than a month later, on November 3rd 1957. This one had the famous dog Laika on board, which soon died of overheating. Both Sputniks were purely scientific, of course, but the USA realised that the Russians would now be able to reach them with intercontinental missiles carrying atomic bombs. The USA did not quite manage to launch a satellite into orbit during the IGY. Eplorer 1, the first American satellite, was launched on February 1st, 1958, still as a part of the IGY programme. In outer space it detected large quantities of electrically charged particles and panic struck again. Could these have originated from secret nuclear high altitude tests by the Russians? But these particles came from solar wind, and were trapped in the earth's magnetic field. They occur in belts around the earth, and have been named Van Allen Belts, after their discoverer James van Allen of Iowa University, leading scientist behind the Explorer Programme. On March 5th 1958, Explorer 2 was launched to further study the radiation belts, but it did not get into space because of rocket failure. On March 25th, Explorer 3 was sent into an eccentric orbit, 186 km from earth at its lowest point (perigee), and 2799 km at the highest point (apogee), specifically to further analyse the Van Allen Belts. The charged particles spiral around magnetic field lines towards one pole, bounce back there and travel all the way to the other pole. Where they get closer to the earth near the poles, they collide with atmospheric molecules, causing the famous auroras. If the particles were not deflected by the magnetic field, radiation on earth would be so strong that life would be impossible.

Nicholas Christofilos, a scientist of Greek origin, came up with a fantastic plan. He postulated that a nuclear explosion in outer space could produce an artificial shield of charged particles, which would be trapped in the earth's magnetic field, and like the Van Allen Belts, would disperse worldwide, from pole to pole. This electro-magnetic shield would fry the guiding electronics of Russian missiles passing through, which would then no longer be able to reach the USA. This is how Operation Argus was born. In the meantime, Russia and the USA had come to an agreement to place a ban on further testing of atomic bombs. A moratorium would probably come into effect by the end of 1958. This meant that the Christofilos experiment had to be done before that, and preparations had to be speeded up.

Operation Argus was approved by the President of the USA on May 1st 1958. On July 26th, Explorer 4 was launched from Cape Canaveral to monitor the effects of Argus in outer space. In August 1958, Task Force 88 sailed for the South Atlantic, with 9 ships and 4500 crew, in deepest secrecy. On August 27th, the first bomb detonated at an altitude of 161 km, at 2:28 am about halfway between Tristan da Cunha and Gough. Had the people lived on the south side of the island, they would have seen a bright flash in the southern sky, and perhaps been exposed to radiation. The crew of the task force probably have been exposed, because at a later age they reported an above average occurrence of leukemia. For the second explosion, the ships moved further south, between Gough Island and Bouvetøya. This one took place at 293 km altitude, at 3:18 am on August 30th. The third and last reached an altitude of 750 km, where it detonated at 22:13 on September 6th. This was the highest nuclear explosion that ever took place in history. The predicted Christofilos effect did occur, but it was not as strong as was hoped and it began to fade away within weeks. As a defence system against Russian missiles it was useless.

Information on the secret plan had already been leaked to the press before it was executed. Two journalists of the *New York Times* knew about it and wanted to go public, but the White House and the Pentagon managed to have them keep it under wraps until at least after the experiment had been done. Scientific papers about the results appeared in early 1959, calling it the greatest scientific achievement since World War II, so it was clear that the secret could not be kept for long. The experiment was declassified, and the *New York Times* published the story on March 19th, 1959. In fact, it is quite surprising that it took almost half a century for the news to reach the people of Tristan da Cunha.

One thing becomes clear when looking into the details of this story: all the planning and preparations for Operation Argus took place after the Gough Island Scientific Survey of 1956. So the Goughalites can sleep peacefully, knowing that they have not been abused for the purpose. That part of the conspiracy theory can go into the rubbish bin. But apart from the question of which authorities, other than the American, knew about the experiments, it remains an outrageous scandal that the people of Tristan da Cunha have never been informed.

44. The latest shipwrecks

In previous chapters I dealt with the numerous shipwrecks on Tristan da Cunha, and the colourful histories about them, up to my first visit in 1993. Since then, I know of four more shipwrecks, all in the 21st century, which brings the total to 36.

In 2005 (I don't know the date), the yacht *Tarion* wrecked off the east side of the harbour. There were two men on board who were brought ashore safely. They remained on the island for a while.

On April 6th 2006 a large towing vessel, the Tug *Mighty Deliverer* left Brazil en route via Cape Town to Singapore. It was towing a Mighty Thing to Deliver indeed, the semi submersible oil platform PXXI. The ship met with very bad weather in the middle of the South Atlantic, not unusual at this time of year. Conditions were so bad that on April 30th the captain of the Tug had to decide to cut the platform loose and let it go, to avoid serious damage to the tug and the towage. The ship stayed near, so if conditions improved the platform could be fastened to the Tug again, to continue its voyage to Singapore. Eye contact with the platform was maintained until 10th May. But swell ruled out getting too close. On 16th May, after a few days with poor visibility, the platform was not seen anymore. The tug tried to find it but failed. The PXXI was declared lost. On 23rd May the Thing was briefly seen again, about 98 miles NW of Tristan. Later that day it was seen 55 miles west of Tristan and 36 miles west of Inaccessible. They lost sight again. A second Tug *Ruby Deliverer* joined in searching for the platform, but with no success. Disaster was about to strike Tristan. Fortunately, the Thing was said to be free of oil and other pollutants…

On 7th June islanders found the missing platform PXXI washed ashore on the south-east coast of Tristan, at Trypot

Beach. The Thing measured 104 metres x 103 meters, and had a draft of 13 metres. It was 6357 gross tonnes in weight. Islanders were on their way to the south coast of the island to collect beef on the first nice day after a very bad, stormy week. To their great surprise they saw the missing platform sitting on the shore in one of Tristan's most inaccessible places. At Trypot, a narrow boulder beach, sheer cliffs rise to about 500 metres. From late August the beach is home to a colony of Northern Rockhopper Penguins. On 8th June the two Police RIBs set out to investigate the platform, and to report to the owners.

They found the Thing marooned on a shallow reef, some 250-300 metres offshore. There was no apparent environmental damage apart from a very small patch of oil.

Meanwhile, the Thing appeared to have changed name. Originally PXXI, in full Petrobras XXI, now officially became known as 'A Turtle', according to its owner, Catleia Oil Company, based in the Cayman Islands. They hired Charles Taylor Consulting, to try and rescue the Turtle. A huge, powerful tug was to be sent from Cape Town to pull the Turtle back into deep water.

On 16th June the salvage tug ST *Zouros Hellas* left Cape Town. It arrived at Tristan on 22nd June, after a very rough passage. It took several days, often again waiting for the weather to improve, to get the rig fully inspected. The salvage team found that on the seaward side the colums were still afloat. This made them very optimistic about the possibility of pulling it free of the reef.

On 29th June the *Zouros Hellas* tried to pull the rig free at high tide. Although the Thing trembled and shook a lot, it did not move one inch. The next month, many attempts were made to gain access to the rig, but the weather was not co-operating in any way: every attempt had to be abandoned prematurely. By 28th July it was decided to pull back the salvage teams and ships, go back to Cape Town, and await better weather in the upcoming spring

311

season. The tugs might return in September to continue their efforts.

On 25th September divers and marine biologists Sue Scott and Geoffrey Fridjohn arrived on the annual SA *Agulhas* voyage. They found many organisms living on the under water parts of the rig that were not native to Tristan waters. Surprisingly enough, they not only found organisms attached to the steel, but also a new species of fish, the South American silver porgy fish *Diplodus argenteus*, freely swimming around the structure in a shoal of 30-50 individuals. They must have survived the extremely rough passage by sticking close to the legs all the time.There was a grave concern that alien species might invade the fragile ecosytem of the Tristan waters, and perhaps might even upset the Tristan Rock Lobster fishery, so vital to the Tristan economy.

During the following months many messages flew accross the world concerning the salvage of the Turtle. In December a special salvage team was flown to Cape Town and shipped to Tristan, more than half a year after the stranding. In January the team filled the leaking parts of the rig with air bags, to push the water out and reduce the weight of the Thing. Also, a lot of scrap metal was removed from the rig to further reduce its weight.

Finally, on 11th February 2007, the tug ST *Titan* managed to pull the structure off the reef, and tow it to the north of the island, where it was scuttled at a depth of 3500 metres, about ten miles east of the Settlement, nine months after the Turtle arrived at Tristan. At last, this turtle was turned.

MS *Oliva* was happily crossing the South Atlantic, nicely staying on the plotted course. Nobody realised that this plotted course ran straight over Nightingale Island. They simply had not looked at the charts properly, and just supposed that the South Atlantic would be an empty space. So, even when Inaccessible Island (which they missed only by inches) appeared on the radar screen, the crew supposed they were looking at a heavy rain shower.

MS *Oliva* ran aground at 4:30 on 16th March 2011, at Spinners Point, the far north-west promontory of Nightingale Island. At 7:00 the fishing vessel MV *Edinburgh* called to Tristan that the ship was well and truly stuck on Nightingale's rocky north coast.

MS *Oliva* was a 75,300 tonne bulk carrier of 225 m length, registered in Malta. The ship was en route from Santos, Brazil, to Singapore, with a load of 65,000 tonnes of soy beans. Can't the Asians grow enough soy beans for themselves, to satisfy their daily requirement of Tahoe, Tofu, and soy sauce? Oliva's operators were based in Greece. The 22 crew consisted of a Greek Captain, and 21 Filipinos. Twelve of them were evacuated to MV *Edinburgh*, shortly after stranding.

The Tristan community, and anyone else in the world who knows about Nightingale as a nature paradise, was very concerned about the environmental risks, especially the chance that rats might escape from the wreck, and swim ashore. Rats on Nightingale would be a total disaster. The Tristan Conservation Department immediately prepared a team to send over to Nightingale to place baited rat traps along the shoreline, opposite the wreck site, although the ship's owner said the ship was certified rat free.

On March 17th a salvage tug left Cape Town, to arrive at Nightingale on the 21st.

On the 17th weather conditions at Nightingale were worsening. It was then decided ro rescue the remaining 10 crew from the *Oliva*, but all day the weather was so bad that rescue was impossible. As the ship was starting to crack and leak the crew on board endured some worrying hours… In the morning another ship arrived on the scene, MV *Prince Albert II*, a cruise ship just visiting Tristan at this dramatic moment. Captain Alexander Golubev immediately offered his assistance, and put his zodiacs and his able and experienced zodiac drivers in the water to get the remaining crew off the ship. That was something different from peacefully landing between penguins and seals with a zodiac full of passengers! Passengers on board the *Prince Albert II* had the most adventurous day of their cruise, watching the zodiacs getting everybody safely on board the Edinburgh by the end of the afternoon. Meanwhile the damaged ship started to leak oil.

During the night, the ship broke up, the heavy swell first breaking the ship's back, and then starting to demolish the superstucture. Oil now flowed freely around the ship. This was a grave situation for the Northern Rockhopper Penguins, of which about 25% of the world population nests on Nightingale. Fortunately, the now full grown chicks had mostly left the island and many adults were also out at sea. But this is the time the adults come back to moult on Nightingale's shores. The ship was carrying about 1500 tonnes of fuel oil.

On the 19th, the salvage tug was about halfway between Cape Town and Tristan, but after arrival, its role would no longer be towing the Oliva into safer waters. They might still be able to pull the larger parts of the wreck off the rocks, and perhaps salvage some of the contents, and then let them sink in deep water, away from the islands. A second ship was sent from Cape Town, chartered by the ship's owners and their insurance company, to assist the Tristan people in cleaning up the oily mess.

On the 21st the salvage tug arrived on the scene. The environmental specialist on board estimated that along Nightingale's shoreline, there were about 20,000 soiled penguins, a gloomy guess, that fortunately would later turn out to be too high. A cleaning-up team was sent over from Tristan, and another team was assembled in Cape Town, ready to travel to Nightingale as well. Oil was now found on neighbouring Inaccessible island too, so the cleaners also went over there to wash penguins. Several hundreds of penguins were taken to Tristan by the Edinburgh, which, after cleaning and feeding, were able to enjoy a clean swim in Tristan's public swimming pool.

On the 26th of March it was decided that the wreckage of the *Oliva* was unsalvagable, and would be left on the rocks to further disintegrate. Meanwile, not only penguins were suffering from oil pollution. In a soiled tidal pool two Fur Seals were found dead, probably poisoned by the oil, while trying to clean themselves. On Inaccessible island oiled seals were also found, and, quite disturbingly, two dead Inaccessible Flightlesss Rails on the beach. They mostly feed on insects, but they may scavenge, like a true small Island Cock, so on the edge of the beaches they might feed on oiled small animals.

On Tristan a Penguin Rehabilitation Centre was established, with a large shed to keep the penguins dry, and where the clean-up of every individual was done. After being cleaned, they were allowed to paddle around in the swimming pool. All penguins had to be fed by hand, daily, an effort not to be underestimated. Penguins were brought in by MV *Edinburgh* on a regular basis, a few hundred at a time. Bad weather prevented penguin transfer on many days. Oiled penguin groups on Nightingale, still awaiting transport, were fenced in, to keep them where they were. By the 4th of April, the total count of penguins brought to Tristan was 3662. Most of these were cleaned successfully. Only 373 died during the process, just about 10%. A survival rate of 90% in such an operation is quite remarkable. From 3rd April onwards the release of fully recovered peguins started.

The crew of MS *Oliva* was picked up by a sister ship, MS *Samaton*, en route to Santos Brazil (to get more soy beans for Singapore?). During the stranding, MS *Oliva* lost one of its life boats. Two years later, in February 2013, it was found, washed ashore on the beach in the Coorong National Park in South Australia, where it is a tourist attraction today. Marine Biologist Sue Scott, who did the survey around the Turtle, came back to Tristan to look at the effects of the stranding of MS Oliva. She found that the thick layer of rotting soy beans on the sea floor caused anaerobic conditions which resulted in massive die-off of marine organisms. Lobster fisheries around Nightingale were suspended for quite a while. There is no indication that any rat ever managed to reach Nightingale alive.

On October 15th 2020, the South African fishing vessel *Geo Searcher* hit a hidden rock near Gough Island and sank. On board were 60 crew members, and two Tristanians, Rodney Green and Ian Lavarello, who were going to do biomass observations around Gough. Only two minor bumps were felt, but when water was seen entering the engine room, and there was no way to stop it, it was evident that sinking was inevitable. In the old days they would have beached the ship to get everybody safely ashore. But Captain Clarence decided otherwise, and sailed away from the island to prevent pollution of the environmentally sensitive coastline with its teeming wildlife. He let his ship sink about one mile north of the island, in 300 m deep water. The *Geo Searcher* was already listing 45 degrees when the last of the crew left the ship. A lifeboat turned over, leaving people swimming for 25 minutes. The four powered fishing boats, which were already in the water, came to assist, and all crew members ended up in life boats and other small craft, which were all joined by ropes, forming one big flotilla. They had to leave all their possessions in the sinking ship. Fortunately, the sea was calm, so they could move around the east coast of the island, where they managed to reach the shore safely near the South African weather station. Station staff helped the men to get ashore, winching up the life rafts. All men were saved and only two had minor injuries. Four days later they were rescued by the South African ship *Agulhas II*, which was sent from Cape Town, chartered by the insurance company. At Tristan, Rodney Green and Ian lavarello were flown ashore by helicopter and the others went back to Cape Town. At Tristan, there were 120 bags with clothes and toiletries waiting for the crew, from 104 households. The people of Tristan da Cunha had reverted to their old habit of helping the victims of shipwrecks.

45. Christmas on Tristan

In 2011 I saw a wonderful voyage to Tristan announced in the Tristan da Cunha Newsletter. A special voyage, with a full week on Tristan around Christmas, specifically catering for members of the Tristan da Cunha Association and the Roland Svenson Foundation in Sweden.

Roland Svenson (1910-2003) was a famous Swedish painter who specialised in withered, deeply grooved faces of hardy fishermen, against a background of dark, gloomy skies. He started his career in what is called 'The Archipelago', the hundreds of small islands east of Stockholm, which are now mostly dotted with holiday cabins, but still had an indiginous population of tough fishermen who had fished there for many generations.

Roland's next destination was the northern Scottish Islands, the Orkneys and the Shetlands, which, with their horrible North Atlantic climate, had even better withered faces. And then he heard about the people of Tristan da Cunha. He read in the newspapers that they had just been evacuated from their island, because of the volcanic eruption in 1961, and were now based in Calshot near Southampton. He could not resist. He visited them there and immediately fell in love with Tristan and its people.

When the Tristanians went back to Tristan in 1963, Roland got permission to join them, and he started to paint withered Tristanian faces against gloomy Tristan skies. This yielded a well-known set of twelve etchings, together with an explanatory booklet. I have the set, of course, and it was not even very expensive (probably meaning he printed too many), and you can still easily find them for sale on the internet, reasonably priced.

Roland was a good Tristan Friend for the rest of his life, and when he died he left a fine sum of his fortune (he had been quite a successful painter) to the Tristan da Cunha Pensioners Fund. His admirers are together in the Roland Svenson Foundation, so like Roland himself, these people are just as crazy about Tristan da Cunha as the members of the Tristan da Cunha Association.

To top it off on this very special voyage the captain of the ship would be Roland's son, Thorvald Svenson. It was evident that I needed to be on this trip, but then I saw the price tag... I was in the the last stage of a bad marriage (after Dineke died I remarried which was a mistake), which meant there was no way I could spend that kind of money on a trip. So I wrote to Noble Caledonia, the British tour operator, enclosing my CV and bluntly told them that they could not do that voyage without me. I did not get a reply from them for many months, so I had given up when I suddenly got the invitation to join as a (paid) guest lecturer. Yes!

This trip was going to be a reverse Odyssey, starting in the Cape Verde Islands. Then, after brief visits to Ascension and St Helena, we would spend a whole week at Tristan da Cunha over Christmas. Then we would go to South Georgia and the Falklands, and finally end in Punta Arenas, on the Chilean side of Strait Magellan (as a British ship, coming from the Falklands you never get permission to sail to Argentinian Ushuaia).

I met passengers and fellow staff members in London, where we would take a special charter flight to Cape Verde. It was like a reunion. There were quite a few passengers I had seen before, either on board Oceanwide's ships, or on the RMS trip, or even both. One of the other staff members was marine biologist Sue Scott, who did the research on the impact of some recent shipwrecks at Tristan and who we had on board the *Professor Molchanov* as a guest. She also brought her husband Michael, a botanist, who had a history of lecturing on cruise ships in the Carribean and on the Amazon. Another interesting guest lecturer was Richard Grundy, who had been a school teacher on Tristan. Expedition leader was Hannah Lawson, distantly related to the ex of famous TV-cook Nigella, whose recipes I liked.

Hannah had inspected the ship a while ago, and

concluded that it was not properly fitted for the Southern Ocean. She found that there were no hand rails in the corridors. These are essential in rough seas, to prevent passengers from falling all over the place. We keep repeating: "always keep one hand for the ship!", but then there must be something to hold on to. So, at her request, the hand rails were installed.

The *Island Sky* was a beautiful ship, but maybe slightly better suited for the Carribean than the Antarctic. It had a flat, low bow which could easily take a lot of water when encountering really big waves, and zodiac operations had to be done from a marina at the stern, instead of a gangway midship, with the risk of much more vertical movements. The voyage was also a bit more luxurious and formal than I was used to. I had to bring a jacket and tie for the captain's welcome reception.

After leaving Cape Verde the first big event was the crossing of the equator. It turned out to be inevitable that I would be King Neptune again. But the ceremony was different. No slapping of bottoms, as the Russians always liked to do, and no goo over the heads of the victims. Instead they had to kneel, humbly ask permission to cross the line, and then kiss the feet (which had been smeared a bit with goo) of my beautiful wife (zodiac driver Guy Esparon from the Seychelles, fitted with huge fake boobs), and finally kiss the fish. For that purpose I had a dead fish on a cord around my neck. With an ominous voice, I would lisp: "kish the fish!" In the course of the day, in the tropical heat, the fish started to liquify but nobody got sick. And, of course, we had a good party afterwards, all throats being desinfected with alcohol.

In Ascension, we made the obligatory trip to the Farm at Green Mountain, admiring the works of Stedson Stroud in his greenhouse, cultivating the native ferns of Ascension. No turtle viewing at night and no zodiac cruise around Boatswain Bird Island. Only a semi-distant view, when we circumnavigated Ascension, which was spectacular anyway, with all the Frigate Birds and Boobies hanging above the ship. In Saint Helena, of course we visited the Napoleon monuments and went to the Millennium Forest, where I had not been before, each of us planting a specimen of an endemic tree. I think I planted a Gumwood.

And then on 23rd December we arrived at Tristan, cloudy as usual. As with the trip with the RMS *Saint Helena*, we were given the opportunity to board with islanders. This time I had booked in advance. I was hosted by Martin and Iris Green, whose marriage I had witnessed in 1993. We had lovely days on Tristan, culminating in Christmas. Michael Scott dressed up as Father Christmas and amused the children, and I was invited to join Christmas dinner at the house of Judy Green, Martin's mother. We had traditional stuffed mutton, and lots of other Tristan delicacies to be scooped up on your plate on a smorgasboard-like buffet.

Then, not unusual at Tristan, but still a bit unexpected in mid summer with calm days and sunshine, the weather changed. A huge, bad system was approaching from the southwest and the captain was very eager to leave Tristan as soon as possible. We would not stay for New Year as was originally planned. On the 26th we were all rounded up and zodiaced to the ship in a rising sea. I was on the last zodiac, and the stern of the ship was already going up and down metres, threatening to crush us when it came down. We all got on board safely, totally soaked, and left Tristan, heading southwest, against the storm.

The centre of this very bad system was lying near South Georgia, and the weather charts showed us that a little further north it was much calmer. Therefore, the captain decided to skip the visit to South Georgia altogether, stay further north and go to the Falkland Islands straight-away instead. A big disappointment for the passengers, but secretly I was looking forward to having more days in the Falklands, where I had never been before. In spite of leaving Tristan early, and skipping the whole South Georgia programme, we only gained two extra days in the Falklands, having to slow down to just a few knots, battling against the wind and the currents all the way in a ship not optimally fitted for the Southern Ocean, also proving that a reverse Atlantic Odyssey is not a good idea. Nevertheless, in the Falklands, we were able to visit four of the western islands, instead of two. We landed on Saunders, Westpoint, New, and Carcass Island, visiting spectacular colonies of Rockhopper Penguins and Black-browed Albatrosses, and saw the endemic Cobb's Wren. When we were approaching Strait Magellan, looking at the time schedule, I was worried about whether we would reach Punta Arenas in time to catch our flights home, but then we were sucked into the First Narrows at an incredible speed of over twenty knots, cruising speed plus an unbelievably strong tidal current, so we made it easily and all got home safely.

46. Stamps

On all my voyages in the Antarctic and the South Atlantic stamps have been an important thing for many passengers. Be it an Antarctic Base, Tristan, or another remote island, it is the same everywhere. Many people forget about roaming around to see interesting things but spend hours in the post office instead, writing and mailing piles of postcards to all their friends and family members. Receiving stamped mail from such a place is a real treat.

For many of our passengers Tristan is the most important place to visit, and quite a few of them are stamp collectors. At the post office they go for the first day covers that are always for sale, and some of them buy piles of them to make other collectors happy in the future, or simply because they are greedy (like the Grabbers on the RMS voyage). Tristan stamps are sought after worldwide, especially the older ones. Most famous are the potato stamps, of course, which had been designed by Allan Crawford.

Cliff, from Canada, was a collector on board. He already had every single stamp of Tristan da Cunha, including the rare potato stamps. But there was still one set he craved, which was even more valuable than all Tristan stamps put together: the so-called Tristan relief stamps, which were issued by Saint Helena for the benefit of the victims of the volcano in 1961. They were just ordinary Saint Helena stamps, but with an overprint, and a little extra value for the Tristan Relief Fund. But then the British Post Office decided that the issue was illegal and ordered all the stamps to be destroyed. They had been for sale for one week, on Saint Helena only. So imagine how rare they are now, and how few people have them. Only 434 sets have been sold. Cliff saw them once for sale on the internet and made a four figure offer but did not get them. I have recently seen them for sale on the website stampworld.com for 5,900 euro, but I am not interested.

Poor Cliff never reached Tristan himself. He was first with us on the Atlantic Odyssey that did not get any further than South Gorgia because of engine trouble. We had to wait for ten days before a relief ship could come and get us. So we had a wonderful opportunity to have all sorts of excursions around Grytviken that we never would have time for otherwise. We were finally rescued and taken to Montevideo by the Argentinian ship *Ushuaia*, a rusty tub whose greatest asset was the fantastic barman Victor, who made the most wonderful Brazilian caipirinhas.

The passengers got a reasonable compensation for the 'unconsumed' part of the voyage, so we saw quite a few of them again in subsequent years. Cliff was back on board two years later, and this time his wife Mary came too. In the Antarctic we had the most abominable weather, with freezing gales straight from the heart of the continent. Remember we are always there in summer when, in the coastal regions that we visit, temperatures are rarely below zero, but this time we had minus 12°C, which meant minus 20-30°C with wind chill. For the first time in my life I watched the sea freeze over with slush forming first which gradually froze into the famous pancake ice. We sought shelter in a bay close to the Argentinian station Esperanza (near Hope Bay, at the tip of the Antarctic Peninsula), waiting for conditions to improve. Suddenly, with the incoming tide, an enormous field of ice floes came around the corner and closed us in, the floes freezing together alarmingly quickly into one solid mass. It was already late autumn and we had visions of spending the winter there, only to be rescued next spring. During the night we had a true blizzard on top of everything, so we had a real Antarctic adventure. Next morning the wind had turned, and wind plus tide gradually drove the ice away. After a few hours we could carefully push our way out. But once out of the ice the rough seas took over, so we had some really bad days on our way to South Georgia. And then Cliff's wife broke her arm when she fell in her cabin. The ship's doctor stabilised the arm, and both of them thought that it would be safe to stay on board for the rest of the trip. But after two days she lost feeling in the arm, indicating that a nerve might be damaged. So, on arrival in South Georgia she was X-rayed in the local hospital. The arm was not just broken but shattered and Mary needed an urgent operation. The nearest port was Port Stanley in the Falkland Islands, so that is where we went. Hundreds of messages went back and forth to insurance companies and many other involved parties. We lost four days on our schedule, so what to do? Eventually we were able to offer a compromise between the various bad solutions: the trip would be prolonged by two days, free of charge, and visits to both Tristan and Saint Helena would be shortened by one day.

In Port Stanley we faced the problem that the authorities would not let the patient ashore as long as there was no guarantee that the ambulance plane, to be sent from Punta Arenas in Chile, was actually on its way. But in the end all went well and Cliff and his wife travelled back to Canada. At this point quite a few other passengers chose to leave because they could not afford to be away from work for two extra days, or simply because they were sick and tired of the Southern Ocean. That also included the couple which had planned to get married on Tristan.

I don't think Cliff will try to get to Tristan for a third time.

Even more valuable than stamps are envelopes that have been sent by expedition leaders, especially from the famous early polar expeditions. When I was preparing my first expedition to Antarctica in 1988 I was approached by Eddy de Busschere from the Belgian Polar Society. He said he was co-ordinating polar mail for collectors worldwide, and asked me if I was willing to take a few pre-addressed envelopes to mail to collectors from Antarctica. I could design an expedition cachet and he would make it for me. I agreed and sent Eddy a drawing of my design (not the obligatory penguin, but a bit more original, with a Sheathbill - one of the species we would study). In return, I received a beautiful rubber cachet, and to my surprise a whole shoebox full of envelopes, not just 'a few'. Normally he would have already glued the postal stamps on the envelopes, but in this case that was not possible. Brazil had an unbelievably high inflation rate in those days, and any stamp bought today would be worthless tomorrow. Instead he gave me sufficient dollars to buy them on the spot. Envelopes were from collectors all over the world, not just Belgium and neighbouring countries. There were envelopes from Australia and New Zealand, and even a few from Brazil.

On Elephant Island Henk and I signed all the envelopes and stamped them with my Sheathbill cachet. On the way home we spent some days at the Brazilian station Comandante Ferraz, where I hoped to mail my envelopes. But there were no stamps for sale anymore as it was end of season. So I took the whole shoebox back home and sent the envelopes back to Eddy, apologising for not having been able to mail them, but at least they had been on Elephant Island and had the expedition stamp and the expedition leader's signature. Hopefully people could be happy with that.

A year later I got a message from Eddy. He had sent the envelopes to his Brazilian co-ordinator, who had managed to get them back to the Brazilian station Comandante Ferraz at the start of the next season. All had been succesfully mailed.

Now envisage this one envelope. No message, just an empty envelope. It was sent by a Brazilian collector to the Brazilian co-ordinator, who sent it to Eddy in Belgium, who sent it to me in the Netherlands. I took it with me to Brazil, then to King George Island in Antarctica, and next to Elephant Island. Then I took it back to King George Island, and back to Brazil, and then back to the Netherlands. Then I sent it back to Eddy in Belgium, who sent it to his Brazilian co-ordinator, who sent it to King George Island, where it was finally mailed to the collector in Brazil. Phew.

47. End of the quest

In 2002 my freezer defrosted. Dineke and I were on holiday in the USA and our daughter was going to use our house for a couple of days. She found the electricity down, probably as a result of a recent thunderstorm which had happened before, and the freezer thawed. Fortunately, it was in a shed outside, otherwise the stench would have permeated every corner of the house. The contents were, of course, lost, except for my two Tristan Moorhens, which I had kept there for almost ten years by this time. They looked, although probably inedible now, still good enough for the skins to be prepared. I took them to the Zoological Museum in Amsterdam, where the skins were indeed successfully prepared. We looked at the stomach contents but found nothing identifiable. No eggshell fragments in any case. I took the skinned bodies back home to re-freeze, but decided that they would be safer in the freezer at my institute.

I also decided that I had waited long enough for the DNA results from the samples Michael Swales collected in 1993. Not a word from him or from Gary Nunn, who would do the analysis. Gary had been completely absorbed by albatross-DNA, which had become his specialty. He is one of the people responsible for splitting up Wandering Albatross populations accross the world into separate (hardly distinguishable) species, and splitting the Yellow-nosed Albatrosses into an Atlantic and an Indian Ocean species. Meanwhile, my institute had also started to embrace DNA-analysis for various purposes, so now I could just ask a colleage to do the job. He was very enthusiastic at first, but after a few months he told me he could not find the additional funds needed or a student do do the actual work. So I went back to Amsterdam, where I had donated my skins. Same response. Enthusiasm first, then a few months silence, and finally the message that they could not fit it into their programme. Next I tried

Wageningen University, with the same results. Now I had been shopping around for more than three years.

Finally, in 2005 there came a solution. I saw a publication of a newly established institute: the Ancient DNA Laboratory at the University of Leiden specialised in analysing DNA samples from old museum specimens. This paper was about the landsnails of the genus *Balea*. These tiny snails, only a few millimeter in length, occur in continental Europe, The British Isles, Iceland, the Azores, and... Tristan da Cunha! DNA-analysis revealed that *Balea* snails from Europe reached the Azores, probably sticking to the feathers of migratory birds, and developed there into two endemic species, one of which was later transported to Madeira, and from there back to Europe, where it now lives side by side with the ancestral species. From the Azores, one species managed to reach Tristan, most likely again with the aid of birds. The authors of this paper thought of long distance migrants like waders, which often accidentally reach islands in the South Atlantic, but in my view Skuas could also be a likely vector. Skuas, originally birds of the far north, managed to reach the Antarctic, through trans-equatorial migration (we see Siberian Pomarine Skuas in the South Atlantic), and establish new species there, one of which later moved back to the North Atlantic, to become the Great Skua of Iceland.

In Tristan da Cunha the *Balea* snails split up into at least eight different species, all endemic to the group, some endemic to only one island, others occurring on two, three, or all four islands, including Gough, indicating that transport by birds must have been a relatively common thing.

I contacted the first author of this interesting paper, professor Edi Gittenberger, and he told me that his study of the land snails of Tristan was greatly stimulated by reading my book on the Moorhen. And, yes, he would be delighted to adopt my Moorhens as a case. So my two corpses moved to Leiden. The museum had plenty of skins from Gough to use, so now we only needed DNA from the type specimen in the British Museum to complete the study. DNA from old bird specimens is usually taken from feathers, which often have tissue remains inside the root of the shaft, especially in feathers which were still growing when the bird was killed. When the skins are treated with arsenic body feathers yield no good DNA. The best results come from flight feathers. But no museum curator in his right mind would allow such a feather to be pulled from a type specimen. Gittenberger and his co-workers had developed another technique. They would drill a miniscule hole in the sole of the foot, to reach untainted tissue, and take a minute sample there, leaving only an almost invisible little hole in the foot. They conferred with their colleagues in Tring, and got them sufficiently interested in the project to obtain permission to take a foot tissue sample from the type specimen of the Tristan Moorhen. So there we go!

First, the DNA of the type specimen differed significantly from the DNA extracted from birds from Gough. Secondly, DNA from both Tristan and Gough seemed to be more different from ordinary Common Moorhens from the Americas than from Eurasia and Africa. Because of the prevailing winds all endemic land birds of Tristan and Gough had been thought to originate from South American relatives. It now seems that the Moorhens came from Africa, and not from South America. Thirdly, the DNA from my corpses was identical to DNA from Gough. That leaves us two firm conclusions. One: the Tristan Moorhen did exist, and differed from the Gough Moorhen (I already had sufficient proof from my early eyewitness accounts, but this genetic confirmation is nice, of course). Two: alas, alas, the present day Island Cocks on Tristan are not survivors from the original stock, but are descendants from the birds that Mister Swales put there (shouldn't 'ave dunnit).

Amazingly, the genetic distance between Tristan and Gough Moorhens seemed to be slightly larger than the distance between Gough moorhens and ordinary Common

Moorhens from Africa. Also, this distance was in the same order of magnitude as the difference between various subspecies of Common Moorhens throughout the world. This suggests that the birds from Tristan and Gough should be seen as one single species, *Gallinua nesiotis*, with two subspecies, *Gallinula nesiotis nesiotis* (extinct) on Tristan, and *Gallinula nesiotis comeri* on Gough. End of story, end of my 40-year quest. The results were published in an online scientific journal (Groenenberg *et al.* 2008).

Interesting personal detail: of the hundreds of papers I have written or co-authored, only the very first and the very last deal with the Moorhens of Tristan da Cunha. This has a symbolic value as if the Moorhen embraces all the other things I have done in my life, like a pair of bookends.

Epilogue

My travels to Antarctica, Tristan da Cunha, and the South Atlantic span a period of almost thirty years. Over the decades I have seen great changes. In the Antarctic, the most obvious change I noted was the colour of the whale bones found ashore in various places. During my first visit in 1988, they were all still beautifully sun bleached white, but over the years their colour changed to brown, grey, or even orange, due to the weather and the action of micro-organisms. For me, the significance of this is the realisation how recent the Great Massacre took place. That it ended in the 1960s had nothing to do with conservation issues. Whaling simply stopped because the stocks were so low that they would no longer sustain a profitable business. On the plus side, we saw a significant increase in the number of whales we encountered during our Atlantic Odysseys, especially in the Humpback Whales and the Fin Whales. We also saw them losing their shyness, and even had curious Fin Whales circling our ship at very close range, playing with us. There has been an increase in the number of ships visiting Antarctica, but that mostly remained unnoticed, since we can now monitor the positions of all ships and carefully try to avoid seeing each other. As a consequence, the rules we have to stick to have become more strict over the years. These rules are self imposed by the International Association of Antarctic Tour Operators (IAATO), and eventually we staff members all had to do an online exam.

In South Georgia we also saw the rules becoming stricter. On my first Odysseys we freely roamed about on Albatross Island, which had a nice population of Wandering Albatrosses, and it was our own responsibility not to disturb the nests of these giants (most wanderers breed on Bird Island, at the westernmost point of South Georgia. Bird Island has a research station, and is off limits for visitors). But later, Albatross Island was also closed for visitors. We could go to nearby Prion Island, which has a smaller population, and there we had to stay on the recently established board walk and viewing platforms, only allowed to go up in small groups at a time. This reduced the experience a bit. Fortunately, the albatrosses sometimes ignored the rules, and walked right up to us, displaying, using the boardwalk themselves. Rules for biosecurity also became stricter, and we now have to carefully vacuum all our clothes, boots, pockets, and bags, to avoid introduction of alien seeds. The greatest change we saw in South Georgia, of course, was the extermination of the reindeer and the rats.

Saint Helena also changed. First we just played with Jonathan on the lawns of Plantation House where the governor lives. Jonathan loved to be scratched under the chin, and of course we did not ride him. But then a Governor's wife decided that the tortoises could only be viewed from a newly established viewing lane, behind a fence. The biggest change in Saint Helena, of course, is the establishment of the airport. We saw that drama develop over the years. Plans went back to the fridge several times. Heavy discussions about the pros and cons. At last, it was constructed, with not a word on environmental compensation, as was promised earlier. The airport took away a significant portion of prime habitat for the endemic, endangered Saint Helena Plover (Wirebird). By sheer coincidence, during one of our visits, the first test flight from Johannesburg arrived, and the pilots vowed never to make that flight again, as the runway is perpendicular (there was no alternative) to the SE trade winds, which can be quite strong here. On the plus side, good progress has been made with the expansion of native flora, like in the Millennium Forest, where Redwoods, Ebony, and the various daisy trees now happily grow together. There is also good news about my favourite animal, the Spiky Yellow Woodlouse, which was critically endangered, believed to be down to less than 100 individuals, hunted down by an invasive, alien spider species. A special project for the Woodlouse was carried out in 2012-2017, expanding the woodland of the Black Cabbage Tree, where the little beasts live in the understorey. With new sampling techniques (they glow when you shine ultraviolet light on them in the dark), the popuation is now thought to be closer to 1000 individuals.

In Ascension, we also had to deal with new rules. At first, we did our own thing at night on the beach with the egg-laying turtles, again leaving it to our own responsibility to behave in a turtle-friendly manner. Some of us would stay on the beach overnight, sleeping in a sheltered hollow (dug by a turtle). I once had the curious experience of waking up because emerging baby turtles crawled along my cheek. Later, we could only visit the turtles with a guide, splitting up into small groups, and we could only witness egg-laying once. On the plus side, the cat extermination in 2005 was a great success. A year later the first boobies from Boatswain Bird Island started to nest on the main island again, and a few years later, the endemic Ascension Frigate Bird followed that example.

And on Tristan? Things have changed too. The people of Tristan have certainly changed into more worldly citizens, much more open to the outside world, now having TV, internet, 24 hours of electricity per day and more contact with visitors. And more children go to the UK for education. The number of inhabitants has dropped from about 300 in the 1990s to around 250 now, mainly due to young people leaving the island for better opportunities elsewhere. The mean age has gone up, and the percentage of pensioners has increased, not only because of departing youngsters, but also because the birth rate has dropped considerably. Families with 12 or more children no longer occur. Women have modernised and became much more emancipated, but they still knit.

Together with the impeccable teeth, the legendary lean bodies of the Tristanians have become history too. Quite a few islanders today carry a fair amount of weight. If you want to reach the age of hundred years with a trim body and a full set of undamaged teeth, the secret is to survive on a near starvation diet of plain potatoes and fish for most of the year, with an occasional rare meat dish, supplemented with abundant eggs and birds in season only. No fruits or veggies, and never brush your teeth.

The asthma project came to a dead end. Noe and Pat went back to Tristan to get more samples, and there have been interesting discussions about ownership of the data in the scientific magazine *Science*. The question was put forward if Tristan da Cunha would be entitled to royalties, when the pharmaceutical industry made millions based on their blood samples. The question remained moot. The genetics of asthma turned out to be too complex, and the magic cure never came. Noe died in 2020, and the people of Tristan da Cunha still cough.

When I went to Tristan in 1993, there were only seven cars on the island, number eight just arriving with the *Agulhas*. Now it seems there are hundreds of them (many dozens anyway - in 2002 fifty were counted). Nobody wants to walk to the Potato Patches anymore and there was a rumour that the road between the Settlement and the Potato Patches needs to be widened to a two-lane road. Rusting cars are a common sight and it may become a problem how to get rid of those.

People have also changed their attitude towards the environment. They don't take birds' eggs and chicks any more on Tristan, and they leave the penguins, seals and albatrosses to reproduce in peace. All wildlife on Tristan is now protected. The biggest achievement, no doubt, was the establishment of a huge sea area around the islands as a totally protected zone on November 13th, 2020. It covers more than 265 thousand square miles, and as such, is the largest protected sea area in the Atlantic, and the fourth largest in the world. The people of Tristan da Cunha have turned into the proud keepers of one of the richest biodiversity hotspots on earth.

On the other hand, although people can find all they need in the supermarket, egging and fatting trips to

Nightingale still occur, much to the dislike of the RSPB. Eggs are no longer being taken by the general public, but are collected by the Conservation Department, which may sound a bit ironic. And the fatting trips have been scaled down considerably. People don't use their famous traditional longboats anymore (what a shame...), so they now have to hire the government launch, limiting the number of participants to twelve only, and also limiting the amount of chicks killed. The Conservation Department is not in favour of continuing this, but the Island Council voted for it. RSPB is not happy again, but in terms of numbers taken the impact is absolutely negligible.

Seal populations have rebounded on Gough, after the massacres in previous centuries, and can be expected to increase on Tristan too. The Silver porgy fish that came with the oil rigg is a coastal, algae grazing species, but nevertheless managed to reach both Inaccessible and Nightingale. There is no indication yet that they have a serious negative impact on the ecosystem.

The mouse eradication programme on Gough has been postponed for many years, because the rat eradication on South Georgia got priority. It was finally scheduled for 2020, and part of the team was already there, but it had to be aborted because of the Covid-19 pandemic. With all the travel restrictions people had great problems getting home. After careful deliberations it had been decided to go on with the project in 2021. Afterwards a camera trap recorded live mice, so the eradication failed; a huge disappointment for everybody.

My first experiences with the people of Tristan da Cunha were not always positive, but on my later visits I saw their friendly side and hospitality, as in the books. It was really nice to meet them in the streets and talk to them in the pub. But there is, of course, a big difference: in 1993 I wanted (needed) to be away from the Settlement as often as possible. On later visits that was never an issue. We sometimes had passengers who wanted to go up the mountain but we would simply tell them that that was not possible. But a little bit of the notorious 'maybe tomorrow' was still there, When approaching the island we always had radio contact first (emails later), with long lists of guided excursions on the island and we had people sign up for these. There was always a walk to the Base on the programme, and there were always people signing up for that, but in reality it almost never took place. The weather was not co-operating, there was no guide available, or... maybe tomorrow. Out of 11 visits on Atlantic Odysseys, the walk to the Base only happened once. We went to Burntwood, where the Base in low and easy to access. We ran up, Tristan speed, were allowed five minutes up there to admire the views of Nightingale and Inaccessible, and then we were chased down again, as fast as possible, to get us back in the safety of the Settlement.

I have now been to Tristan 14 times: 11 Odysseys, my first trip with the *Agulhas*, the quincentenary trip with the RMS, and the voyage with the *Island Sky*. Out of these 14, we were not able to land on four occasions, three times because of bad weather, and once because there was a virus on the island, and we could not risk being thrown into quarantaine on our next destination (Saint Helena). I have been lucky enough to see most of the things that can be seen on and around Tristan da Cunha. I have circumnavigated all the islands multiple times, landed on Nightingale many times, landed on Inaccessible Island once, and saw the Holy (G)rail. I did not land on Gough, but touched the rocks from the zodiac on a flat calm day. I saw all the birds and sea mammals that can be seen, and on Tristan I have been in many houses, slept in three of them, and had a full Christmas dinner. I climbed the new volcano and the Hillpiece, went to the Potato Patches, stayed overnight at Sandy Point and the Caves, and saw the Ponds, the Molly study plot, and the crater lake in the Peak. What else is there to wish for?

In the Settlement, I always tried to get a lobster sandwich, and always checked in the museum if my wooden shoes were still there. I never failed to visit Joyce Hagan, my oldest friend on Tristan, and have a cup of tea or a beer with her, and I say hello to Connie Glass and a few others. And I always see Jeff, and share a generous shot of vodka, regardless of the time of day. Jeff likes to remind me that I was drunk ('touched-up') when we were at the Caves together in 1993.

And the Island Cocks? The people still hate them. On one of my visits, well educated, three times Chief Islander James Glass asked me if I was going after these birds again, and added: "You can kill as many as you want."

Can I convince the people that there were Flightless Moorhens on their island, two centuries ago? I doubt it. But for myself, and for Science in general, the conclusion is crystal clear: the now extinct Flightless Moorhen of Tristan da Cunha was very, very real.

Acknowledgements

First of all, I have to thank my late wife, Dineke, who always supported my crazy ideas, and who was happily willing to spend a whole year with me on Nightingale and another one on Tristan. In hindsight, we may have been lucky it never happened. I am not so sure our marriage would have survived it. Having seen the terrain in both islands, the kind of fieldwork we had in mind would have been virtually impossible. Dineke also critically went through the manuscript of the Dutch version of my book in 1997, and my daughter Nienke Beintema carefully checked the grammar and spelling.

Other people who helped me in those days were, in alphabetical order, Mr C.W. Benson, Marietje Brander, John Cooper, Allan Crawford, Hugh Elliott, Mike Frazer, Conrad (Connie) Glass, Herbert Glass, James (Jimmy) Glass, Lewis Glass, Barton Green, Harold Green, Iris Green, Judy Green, Martin Green, Nelson Green, Winnie Green, Wim Groeneweg, Joyce Hagan, Patrick Helyer, Martin Holdgate, Philip Johnson, Marnix Koolhaas, Botte Kuiper, Liesbeth Oskamp, André Repetto, Jeff Rogers, Peter Ryan, Janneke Schokker, Neil Swain, Michael Swales, Albert Veldkamp, Nigel Wace, Jan Wattel, and Noe Zamel.

Of these people, islander Jeff Rogers takes a special place, taking me to Sandy Point and the Caves, and I especially have to mention his dog Number, who enthusiastically secured the all important specimens of Island Cocks for me.

For the present version of this book, I had help from more people. First I have to thank Edi Gittenberger and Dick Groenenberg, Leiden University, for doing the final DNA analysis of the Moorhens of Gough and Tristan da Cunha.

In the history department, I was generously helped by José Manuel Malhão Pereira from Portugal, who supplied me with piles of information on the old Portuguese voyages in Tristan waters. Martin Timoney, Irish archeologist, who studies the life of the Smith sisters, supplied me with some details about these girls.

Others who helped me in finding new (old) sources (sometimes in vain), were Richard Grundy, Sandra Kornet-Duyvenboden, Ko de Korte, Pedro Lourenço, Peter Millington, Sarah Sanders, Irene Schaffer, Menno Schilthuizen, and Sue Scott.

Visiting Tristan so often, has been made possible by tour operators Oceanwide Expeditions and Noble Caledonia, who allowed me to lecture on their ships.

Last, but cerainly not least, there is Sue McVerry, who painstakingly corrected and improved my English. Sue has been to Tristan da Cunha once, on an Atlantic Odyssey years ago. More recently she saw one of my books (about islands, in Dutch) for sale in a charity shop in Australia. She contacted me, and, being a writer herself, offered to edit my texts, if I ever contemplated publishing a book in English. So there we are.

Appendix 1

Birds of the Tristan da Cunha archipelago

T = Tristan, N = Nightingale, I = Inaccessible, G = Gough,
? = breeding not certain, E = extinct.

Scientific name	English name	local Tristan name	island
Eudyptes moseleyi	Northern Rockhopper	Pinnamin	T, N, I, G
Diomedea dabbenena	Tristan Albatross	Gony	I, G
Thalassarche chlororhyncos	Atlantic Yellow-nosed Albatross	Molly	T, N, I, G
Phoebetria fusca	Sooty Albatross	Peeo	T, N, I, G
Macronectes giganteus	Southern Giant Petrel	Stinker	G
Pterodroma macroptera	Great-winged Petrel	Black Haglet	T, I?, G
Pterodroma incerta	Atlantic Petrel	Biggest Whitebreast	T, I?, G
Pterodroma brevirostris	Kerguelen Petrel	Blue Nighthawk	T, I?, G
Pterodroma mollis	Soft-plumaged Petrel	Littlest Whitebreast	T, N, I, G
Pachyptila vittata	Broad-billed Prion	Nightbird	T, N, I, G
Procellaria conspicillata	Spectabled Petrel	Ringeye	I
Procellaria cinerea	Grey Petrel	Pediunker	T, I?, G
Puffinus assimilis	Little Shearwater	Whistler	T?, N, I, G
Puffinus gravis	Greater Shearwater	Petrel	T?, N, I, G
Puffinus griseus	Sooty Shearwater	-	T
Garrodia nereis	Grey-backed Storm Petrel	-	G
Pelagodroma marina	White-faced Storm Petrel	Skipjack	T?, N, I, G
Fregetta grallaria	White-bellied Storm Petrel	Storm Pigeon	N, I, G
Pelecanoides urinatrix	Common Diving Petrel	Flying Pinnamin	N, I, G
Stercorarius antarcticus	Brown Skua	Seahen	T, N, I, G
Sterna vittata	Antarctic Tern	Kingbird	T, N, I, G
Anous stolidus	Brown Noddy	Wood Pigeon	T, N, I, G
Gallinula nesiotis	Tristan Moorhen/Gough Moorhen	Island Cock	T(E), G
Atlantisia rogersi	Inaccessible Flightless Rail	Little Island Cock	I
Nesocichla eremita	Tristan Thrush	Starchy	T, N, I
Nesospiza wilkinsi	Wilkins' Bunting	Big Canary	N, I
Nesospiza questi	Nightingale Bunting	Canary	N
Nesospiza acunhae	Tristan Bunting	Canary	T(E), I
Rowettia goughensis	Gough Bunting	Bunting	G

Appendix 2

Shipping prior to 1800

Po = Portugal, Nl = Netherlands, Br = Britain, Fr = France, Au = Austria, US = United States of America. Visitors to Tristan may also have been to Gough. Towards the end of the 18th Century, there also have been many unrecorded visits of whalers and sealers, often from the US. Numbers in the column 'source' refer to: 1. Brander (1940,1952), 2. Bruijn et al. (1979), 3. Gray Birch (1875), 4. Headland (1992), 5. Coolhaas (1964), 6. Milner & Brierly (1869), 7. Oliver (1891), 8. Smith (1991), 9. D'Après de Mannevillette (1775), 10. Stapel (1927), 11. Faustini (undated), 12. Purdy (1816), 13. Malhão Pereira (2001), 14. Soeiro de Brito et al (1992).

Year	ships and captains	from	to	source
1505	Gonçalo Alvares	Po	Gough	4
1506	*Espirito Santo*, Tristão da Cunha			
	Flor de la Mar, Afonso de Albuquerque	Po	Tristan	1,3
1508	*Sam João*, Jorge de Aguiar (shipwrecked!)	Po	Tristan	14
1520	*Sam Rafael* Rui Vaz Pereira	Po	Tristan?	11
1535	*Espera*, Fernão Peres de Andrade	Po	Tristan	13
1537	André Vaz	Po	Tristan?	13
1557	*Santa Maria da Graça*,			
	Luis Fernandes Vasconcelos	Po	Tristan?	11
1583	unnamed fleet	Po	Tristan?	8
1601	*Bruinvis*, Willem van Westzanen	Nl	Tristan	1
1620	*Montmorancy, Esperance*, and *Hermitage*,			
	Beaulieu	Fr	Tristan	11
1629	*Hollandia, Der Goes, Westzaenen*,			
	Oostzaenen, Jacob Speckx	Nl	Tristan	5
1630	*Deventer, Middelburg, Hof van Holland*,			
	Artus Gijsels	Nl	Tristan	5
1643	*Heemstede*, Claes Gerritsz. Bierenbroodspot Nl		Tristan	1
1656	*'t Nachtglas*, Jan Jacobsz.	Nl	Tristan	1
1657	*Orangie*, Rijklof van Goens	Nl	Gough	10
1659	*Graveland*,	Nl	Tristan	1
1666	Marquis de Mondevergne	Fr	Tristan	11
1669	*Grundel*, Gerritz. Riddermuis	Nl	Tristan	1
1675	Antoine de la Roche	Br	Gough?	4
1676	*Vautour*	Fr	Tristan	4
1681	*Ternate*, Gerritz.	Nl	Tristan	2
1685	*Welfare, Kent, Rainbow*	Br	Tristan	4
1690	*Hirondelle*, Valleau, François Leguat	Nl	Tristan	4,7
1696	*Geelvinck, Nijptang, t'Wezeltje*,			
	Willem de Vlaming	Nl	Tristan	1

1700	*Paramore*, Edmund Halley	Br	Tristan	4
1708	*St. Louis*	Fr	Tristan?	4
1711	*l'Adelaide, l'Eclatant, le Fendant,*			
	Housfaije, le Chevalier	Fr	Tristan	9
1732	*Richmond*, Gough	Br	Gough	4
1755	*Le Rouillé*, Joran	Fr	Tristan	9,11
1758	*Osterly*, Vincent	Br	Gough	4
1760	Gamiel Nightingale	Br	Tristan	1
1767	*Etoile du Matin, l'Heure du Berger,*			
	d'Etchevery, Sieur Donat	Fr	Tristan	1
1775	*Joseph et Thérèse*, Bolts	Au	Tristan	1
1777	*Cormorant, Rippon*, Vernon	Br	Tristan	4
1790	*Betsy*, Colquhoun	US	Tristan	6
1790	*Industry*, Patten	US	Tristan	1
1791	*Philadelphia*, Cahoone	US	Tristan	4
1791	*Warren*, Smith	US	Tristan	4
1792	*Grand Turk*, Hodges	US	Tristan	4
1792	*Lion, Hindostan, Jackal*, Gower	Br	Tristan	1,4
1792	*General Elliott*	Br	Tristan	8
1793	*Le Courier*, Le Gars	Fr	Tristan	1,4
1794	*Essex*	Br	Tristan	8
1795	*Providence*, Broughton	Br?	Tristan	12
1799	*Sally*, Péron	Fr	Tristan	4

For clarity, below I name the ships/captains, which have been listed by various authors, but to my best judgment never saw the islands (or did not even exist: *Capitão Mor* and *Vlaming*):

1503 De Gonneville

1506 *Capitão Mor*

1598 Van Neck

1610 *Globe*

1618 (1619) Bontekoe

1626 Dutch fleet

1646 *Witte Olifant*

1646 *Koning David*

1646 *Witte Paard*

1681 *Elburg*

1684 *Tonquin Merchant*

1697 *Vlaming*

1775 D'Après de Mannevillette

Appendix 3

Shipwrecks of Tristan da Cunha

T = Tristan, I = Inaccessible, N = Nightingale, G = Gough

1.	1508	*Sam João*	T
2.	1817	*Julia*	T
3.	1820	*Sarah*	T
4.	1821	*Blenden Hall*	I
5.	1825	*Nassau*	T
6.	1836	*Emily*	T
7.	1856	*Joseph Somes*	S
8.	1864	*Lark*	T
9.	1869	*Bogota*	S
10.	1870	*Sir Ralph Abercombie*	T
11.	1871	*Beacon Light*	S
12.	1872	*Czarina*	T
13.	1872	*Olympia*	S
14.	1878	*Mabel Clark*	T
15.	1878	*Philena Winslow*	G
16.	1881	*Edward Vittory*	T
17.	1882	*Henry B. Paul*	T
18.	1883	*Shakespeare*	I
19.	1885	life-boat disaster	S
20.	1892	*Italia*	T
21.	1893	*Allenshaw*	T
22.	1897	*Helen St. Lea*	I
23.	1898	*Glenhuntly*	S
24.	1928	*København*	S
25.	1953	*Coimbra*	T
26.	1971	*Mount Alkis*	S
27.	1986	*Chricanto*	S
28.	1987	*Brandgans*	T
29.	1988	*Aku Aku*	T
30.	1993	*Halcyon*	T
31.	1993	unnamed	I (perhaps the same as 30)
32.	1993	*Polessk*	S
33.	2005	*Tarion*	T
34.	2006	*Oliva*	N
35.	2011	*Mighty Deliverer* (the 'Turtle')	T
36.	2020	*Geo Searcher*	G

Appendix 4

Observations of the extinct Island Cock of Tristan da Cunha

17?? - It is possible that the 'strange bird that goes upright' mentioned by Dalrymple was a Moorhen, resembling August Earle's description ' their gait was something like that of the penguin.' Dalrymple quoted from the English Pilot, which was published in many editions in the 17th and 18th century. I have no idea how far back this decription goes.

1790 - Th first man to spend a long time on Tristan, Patten, saw and ate Moorhens (Morrell 1832, Purdy 1816).

1811 - King Jonathan saw and ate Moorhens (Im Thurn and Wharton 1925).

1816 - Captain Dugald Carmichael saw and collected Moorhens (Carmichael 1818). Three were sold at the Bullock auction sale (Bullock 1819), but afterwards were never found (Stresemann 1953).

1824 - August Earle's dog occasionally caught tasty 'Black partridges'. Jules Verne, in his *The Children of Captain Grant*, relates a visit to Tristan, mentioning the history of the island, including Earle's involuntary stay. In his book, Lord Glenarvan goes ashore to shoot several couples of Black partridges, from which the cook made an excellent dish (Verne 1867-1868).

1835 - William Stirling, on board the *Tiger*, visited Tristan and saw Moorhens (Stirling 1843).

1842 - Brierly, travelling from Plymouth to Australia, took two Moorhens on board (Brierly 1842, Wace & Holdgate 1976). These have never been found.

1852 - Macgillivray, on board the *Herald*, stayed three days on Tristan. He found the Moorhens to still be common, especially away from the Settlement (Macgillivray 1852, Wace & holdgate 1976).

1853 - Mr. Gurney received a letter from Mr Strange, from Sydney. Strange writes that he had met someone who saw birds on Tristan 'without wings' (Gurney 1853). We find a bird without wings in the boy's book *A grue of Ice* by Jenkins (1962), which in a heavy gale was blown from the island onto the ship, where the author takes the lack of wings a bit too literally.

1856 - Captain Nolloth of the *Frolic* reported that the Moorhens on Tristan, according to the islanders, were still common. He also said they considered them a great delicacy. Nolloth collected geological and biological specimens, including a Moorhen egg (Nolloth 1856). The egg is supposed to be in the Cape Town museum, but cannot be found.

1861 - Moorhens were taken to London and were desribed as a new species by Sclater (Sclater 1861).

1867 - During the Royal visit with the *Galatea* (when the Settlement was named after the Duke of Edinburgh). the islanders said that flightless Cocks were no longer to be found near the Settlement, but were still common further away (Milner & Brierly 1869).

1868 - Sperling visited Tristan and reported that the Moorhens had become scarce (Sperling 1872).

1869 - Mr Layard in Cape Town received a collection of eggs and birds from 'islands in the neighbourhood of Tristan da Cunha', including Moorhens. Three were sent to London (Layard 1869). Two of those got lost, the third probably ended up in the American Museum of Natural History in Washington in 1871. In this case, these birds were probably Gough Moorhens (Beintema 1972).

1872 - The islanders sold a Moorhen to a ship. No further information (Willemoes Suhm) 1876).

1873 - The *Challenger* expedition failed to find Moorhens, but did not venture further than half an hour walking distance from the Settlement (Moseley 1879, Thomson 1877).

1888 - George Comer collected Moorhens on Gough, which were described as a new species (Allen 1892).

1906 - Nicoll, naturalist on board the *Valhalla*, visited Tristan. The islanders told him Moorhens were no longer to be found (Nicoll 1906).

1937 - Yngvar Hagen, member of the Norwegian expedition, could not find anyone who had ever seen a Moorhen (Hagen 1952).

1950 - Hugh Elliott, first Administrator of Tristan, tried again, but even the oldest people on the island had never seen one and could not even remember anyone who had (Elliott 1957).

1981 - In the boy's book *The Islanders*, where the author was inspired by Tristan history, the people knew about the former existence of flightless 'Island Fowl', which were exterminated by their ancestors, because it was so easy to catch them (Townsend 1981).

References

Too much has been written about Tristan da Cunha to sum up here. Helyer & Swales (1998) assembled thousands of titles, and even they are not complete. Below, I limit myself to books, mainly, except for the sources about Moorhens. I have four categories:

1. The real Tristan books and publications
2. Exploration, travel and island books
3. Fiction
4. Moorhen sources

1. **The real Tristan books and publications**

Barrow, K.M. 1910. *Three years in Tristan da Cunha*. Skeffington, London.

Beintema, A.J. 1997. *Het waterhoentje van Tristan da Cunha*. Atlas, Amsterdam.

Blair, J.P. 1964. Home to Lonely Tristan da Cunha. *Nat. Geogr. Mag.* **125** (1): 60-81.

Booy, D.M. 1957. *Rock of Exile: a Narrative of Tristan da Cunha*. Dent & Sons, London.

Brander, J. 1940. *Tristan da Cunha, 1506-1902*. Allen & Unwin, London.

Brander, J. 1952. *Tristan da Cunha, 1506-1950. Geschiedenis van een volkplanting*. Uitgeversmaatschappij West-Friesland, Hoorn.

Butler, H.J. 1952. *Ships calling at Tristan da Cunha* 1506-1952. Private publication (not found!).

Buxton, E. 2001. Island *Chaplain Tristan da Cunha 1975-1978*. George Mann Publications, Easton, Winchester.

Christophersen, E. 1938. *Tristan da Cunha: den ensommy øy*. Ashehaug & Co, Oslo. English translation: *Tristan da Cunha - the Lonely Isle*. Cassell, London, 1940.

Christophersen, E. (ed.) et al. 1940-1968. *Results of the Norwegian Scientific Expedition to Tristan da Cunha, 1937-38*, nos 1-55, 3 vols. Det Norske Videnskaps Akademi, Oslo.

Cooper, J. & P.G. Ryan, 1993. *Management Plan for the Gough Island Wildlife Reserve*. Goverment of Tristan da Cunha, Tristan da Cunha.

Crabb, G. 1980. *The History and Postal History of Tristan da Cunha*. Private publication.

Crawford, A.B. 1941. *I went to Tristan*. Hodder & Stoughton, London.

Crawford, A.B. 1982. *Tristan da Cunha and the Roaring Fourties*. Charles Skilton Ltd, Edinburgh, London. David Philip, Cape Town.

Crawford, A.B. 1999. *Penguins, Potatoes & Postage Stamps*. Anthony Nelson, Oswestry.

Crawford, A.B. 2004. *Tristan da Cunha: Wartime Invasion*. George Mann Publications, Easton, Winchester.

Crawford A.B. undated (2004).*Tristan da Cunha. Glimpses into Past History*. Private publication for the benefit of the Tristan da Cunha Pensioners Fund.

Crovari, J. 1990. *Tristan da Cunha, l'Isola delle aragoste*. Silver Press Genova.

Earle, A. 1833. *Narrative of a Nine Month's residence in New Zealand in 1827 together with a Journal of Residence in Tristan da Cunha, an Island situated between South America and the Cape of Good Hope*. Longman, London.

Falk-Rønne, A. 1963. *Tilbage til Tristan*. Pondus-Bøgerne, Lohse. English translation: *Back to Tristan*. Allen & Unwin, London.

Faustini, A. undated (1925?). *The Annals of Tristan da Cunha*. Online: http://www.tristan.it/TRISTAN/tristanlibri/tristan_anna ls.pdf.

Foran, W.R. 1938. Tristan da Cunha, Isles of Contenment. *Nat. Geogr. Mag.* **74** (5): 671-694.

Frazer, M., D. Gilfillan, N. Hall, R. Holt, N. Mateer, R. Preece, C. Siddall, M. Swales, J. Wooley & D. Dowsett, 1983. *Denstone Expedition to Inaccessible Island*. Denstonian supplement. Tresises, Burton upon Trent & Derby.

Gane, D.M. 1932. *Tristan da Cunha. An Empire Outpost and its Keepers with Glimpses of its Past and Consideration of the Future*. Allen & Unwin, London.

Glass, C.J. 2005. *Rockhopper Copper*. Orphans Press, Leominster.

Glass, J. & A. Green, 2003. *A short Guide to Tristan da Cunha*. Whitby Press, Whitby.

Greig, A.M. 1847. *Fate of the 'Blenden Hall', East Indiaman, Captain Alexander Greig, bound for Bombay: with an Account of the Wreck and Sufferings and Privations endured by the Survivors, for Six Months on the Desolate Islands of Inaccessible and Tristan d'Acunha*. William H. Colyer, New York.

Hagen, Y. 1952. *Birds of Tristan da Cunha*. Results of the Norwegian Scientific Expedition to Tristan da Cunha, 1937-38, no 20. Det Norske Videnskaps Akademi, Oslo.

Helyer, P. & M.K. Swales, 1998. *Bibliography of Tristan da Cunha*. Anthony Nelson, Oswestry.

Holdgate, M.W. 1958. *Mountains in the Sea*. Macmillan, London.

Hosegood, N. 1964. *The Glass Island. The Story of Tristan da Cunha*. Hodder & Stoughton, London.

Kimber, G. 2020. *Between the mountains and the sea. Memories of a childhood on Tristan da Cunha, the world's loneliest inhabited island*. Private publication.

Kornet-van Duyvenboden, S. 2004. *Op zoek naar Pieter Groen*. Boekhandel Van den Berg, Katwijk aan de Rijn. English translation: *A Dutchman on Tristan da Cunha. The quest for Peter Green*. George Mann Publications, Easton, Winchester, 2007.

Lajolo, A. & G. Lombardi, 1994. *L'Isola in capo al Mondo*. Nuova Eri Edizioni Rai, Torino.

Lajolo, A. & G. Lombardi, 1999. *Tristan da Cunha l'isola leggendaria; Tristan da Cunha the Legendary Island*. Museo Marinaro Tommasino, Andreatta.

Lavarello, A. 1936. *I naufragi di Tristan*. Istituto Editoriale Avio Navale, Milano.

Lavarello, P. undated. *Tristan da Cunha Recipe Book*. Mothers Union St. Mary's Church, Tristan da Cunha.

Lewis, L. 1950. New Life for the 'Loneliest Isle'. *Nat. Geogr. Mag.* **97** (1): 105-116.

Ljung, K. & S. Björck, 2011. A pollen record of the last 450 years from a lowland peat bog on Tristan da Cunha, South Atlantic, implying early anthropogenic influence, *Journal of Quaterly Science*, July 2011. Online: https://onlinelibrary.wiley.com/doi/abs/10.1002/jqs.1489.

Lockhart, J.G. 1930. *'Blenden Hall'; the True Story of a Shipwreck; a casting Away and Life on a Desert Island*. Philip & Allen, London.

McCormick, E.H. (ed.) 1966. *Narrative of a Residence in New Zealand and a Journal of Residence in Tristan da Cunha by Augustus Earle*. Clarendon Press, Oxford.

McKay, M. 1963. *The Angry Island: the Story of Tristan da Cunha (1506-1963)*. Arthur Barker, London.

Munch, P.A. 1945. *Sociology of Tristan da Cunha*. Results of the Norwegian Scientific Expedition to Tristan da Cunha, 1937-38, no 13. Det Norske Videnskaps Akademi, Oslo.

Munch. P.A. 1970. *The Song Tradition of Tristan da Cunha*. Indiana University Research Center for the Language Sciences, Bloomington; Mouton & Co., Den Haag.

Munch, P.A. 1971. *Crisis in Utopia*. Longmans, London.

Munch, P.A. 2008. *Glimpsing Utopia: Tristan da Cunha 1937-38, A Norwegian Diary*. Translated and edited by Catherine Munch Snyder. George Mann Publications, Easton, Winchester.

Rogers, R.A. 1927. *The Lonely Island*, Allen & Unwin, London.

Rosenthal, E. 1952. *Shelter from the Spray*. Timmins, Cape Town.

Ryan, P. (ed.) 2007. *Field Guide to the animals and plants of Tristan da Cunha and Gough Island*. Pisces Publications, Newbury.

Schreier, D. 2003. *Isolation and Language Change. Contemporary and Sociohistorical Evidence from Tristan da Cunha English*. Palgrave MacMillan, Houndmills, Basingstoke.

Schreier, D. & K. Lavarello-Schreier, 2003. *Tristan da Cunha. History, People, Language*. Battlebridge Publications, London.

Smith, R.C. 1986. *The Ships of Tristan da Cunha.* Vol. 1. *Lifeline of a Lonely Island.* Private publication.

Smith, R.C. 1991. *The ships of Tristan da Cunha. A listing 1506-1991.* Private Publication.

Smith, R.C. 1994. *The Ships of Tristan da Cunha.* Vol. 2. *A history and the mails.* Private publication.

Swales, M.K., C. Whirledge, J. Wooley & S. Wright 1993. *Denstone Expedition to Tristan da Cunha.* Denstonian supplement. Tresises, Burton upon Trent & Derby.

Soodyall, H., T. Jenkins, A, Mukherjee, E. Dutoit, D.F. Roberts & M. Stoneking, 1997. The founding mitochondrial DNA lineages of Tristan da Cunha islanders. *Am. J. Phys. Anthropol.* **104**: 157-166. Online (abstract only): https://onlinelibrary.wiley.com/doi/10.1002/(SICI)1096-8644(199710)104:2%3C157::AID-AJPA2%3E3.0.CO;2-W.

Soodyall, H., A. Nebel, B. Morar & T. Jenkins, 2003. Genealogy and genes: tracing the founding fathers of Tristan da Cunha. *European Journal of human Genetics* **11**: 705-709. Online: https://www.nature.com/articles/5201022.

Svensson, R. 1965. *Tristan da Cunha, South Atlantic.* Propius Publishing House, Stockholm.

Taylor, R.W. 2001-2008. *Tristan da Cunha Monographs.* 18 vols. Private publication.

Taylor, W.F. 1856. *Some Account of the Settlement of Tristan d'Acunha, in the South Atlantic Ocean.* Society for Promoting Christian Knowledge, London.

Wace, N.M. & M.W. Holdgate 1976. *Man and Nature in the Tristan da Cunha Islands.* IUCN Monograph No. 6. IUCN, Morges.

Wheeler, P.J.F. 1962. Death of an Island. *Nat. Geogr. Mag.* **121** (5): 678-695.

Wooley, J. 1994. *Deliver me from Safety*. Wilton 65, York.

Zettersten, A. 1969. *The English of Tristan da Cunha*. Gleering, Lund.

2. **Exploration, travel and island books**

Alexander, J. 1982. Voices *and Echoes. Tales from Colonial Women*. Quartet Books Ltd., London.

Andersen, L. 1935. *Lis sails the Atlantic*. George Routledge & Sons Ltd., London.

Baarslag, K. 1941. *Islands of Adventure*. Robert Hale Ltd., London.

Baily, L. (ed.) 1945. *Travellers' Tales*. Allen & Unwin, London.

Barlow, J. 1806. *A Voyage to Cochinchina in the Years 1792 and 1793*. Cadell & Davies, London. Republished by Oxford University Press, 1975.

Baty, R. du 1910. *15,000 miles in a Ketch*. Nelson & Sons, London.

Bruijn, J.R., F.S. Gaastra & I. Schöffer, 1979. *Dutch Asiatic Shipping*. Martinus Nijhoff, The Hague.

Bull. H.J. 1896. *Cruise of the Antarctic*. Edward Arnold, London.

Bullen, F. 1898. *The Cruise of the Cachalot*. F.M. Lupton Pub. Co. Reprinted by various publishers many times, up to 2008.

Burgess, A. 1941. *No Risks - no Romance*. Jonathan Cape, London.

Cannan, E. 1992. *Churches in the South Atlantic Islands, 1502-1991*. Anthony Nelson, Oswestry.

Christensen, L. 1935. *Such is the Antarctic.* Hodder & Stoughton, London.

Coolhaas, W. Ph. 1960-1975. *Generale missiven van Gouverneurs- Generaal en Raden aan Heren XVII der Verenigde Oostindische Compagnie.* 5 vols. Martinus Nijhoff, The Hague.

Cooper, G. 1960. *Isles of Romance and Mystery.* Publ. Latherwell Press.

Crawford, A.B. 2006. *Memoirs: Noth, South, East & West.* George Mann Publications, Easton, Winchester.

Croix, T. de la 1960. *Mysteries of the Islands.* Frederick Muller Ltd., London.

Dalrymple, A. 1786. *A serious Admonition to the Publick on the Intended Thief-Colony at Botany Bay.* George Bigg, London. Online: https://nla.gov.au/nla.obj-19515009/view?partId=nla.obj-19595765#page/n6/mode/1up

D'Après de Mannevillette 1745. *Le Neptune Oriental ou Routier Général des Côtes des Indes Orientales et de la Chine enrichi de cartes hydrographiques.* Jean François Routel, Paris. Reprinted in 1775 by Demonville, Paris.

Drower, G. 1992. *Britain's Dependent Territories: a Fistful of Islands.* Dartmouth, London.

Evans, D. 1994. *Schooling in the South Atlantic Islands, 1961-1992.* Anthony Nelson, Oswestry.

Fogle, B. 2003. *The Teatime Islands. Journeys to Britain's Faraway Outposts.* Michael Joseph, Penguin Books, London.

Fontoura da Costa, A. 1940. *Roteiros inéditos da carreira da India do século XVI/ Prefacianados e anotados.* Ed. comemorativa do Duplo Centenário da Fundacão e Restauração de Portugal. Agência Geral das Colónias, Lisboa.

Fontoura da Costa, A. 1960. *A marinharia dos descobrimentos*. Agencia Geral do Ultramar, Lisboa.

Gray, Bishop 1856. *Three Months' Visitation by the Bishop of Cape Town, in the Autumn of 1855: with an Account of his Voyage to the Island of Tristan d'Acunha, in March, 1856*. Bell and Daldy, London.

Gray Birch, W. de 1875. *Commentaries of the Great Afonso do Alboquerque, Second Viceroy of India*. Hakluyt Society, Vol 53, part 1. London. Reprint by Elibron Classics, 2005.

Green, L.G. undated (before 1962). *Where Man still dream*. Timmins, Cape Town.

Green, L.G. undated, reprint 1982. *A Decent Fellow doesn't work*. Timmins, Cape Town.

Green, L.G. 1960. *Eight Bells at Salamander*. Timmins, Cape Town.

Green, L.G. 1962. *Islands Time forgot*. Putnam, London.

Green, L.G. 1965. *Almost forgotten, Never told*. Timmins, Cape Town.

Green, L.G. 1972. *When the Journey is over*. Timmins, Cape Town.

Hackford-Jones, J. 1980. *Augustus Earle, Travel Artist. Paintings and Drawings in the Rex Nan Kivell Collection, National Library of Australia*. Scolar Press, London/ national Library of Australia, Canberra.

Hardy, A.C. 1967. *Great Waters; a Voyage of Natural History to study Whales, Plankton, and Waters of the Southern Ocean in the Old Royal Research Ship 'Discovery' with the Results brought up to Date by the findings of the R.R.S. 'Discovery II'*. Collins, London.

Harwood, D. 1969. *Alert to Danger*. Bell, London.

Headland, R.K. 1992. *Chronological List of Antarctic Expeditions and Related Historical Events*. Cambridge University Press, Cambridge.

His Royal Highness The Duke of Edinburgh, 1962. *Birds from Brittannia*. Longmans, London.

Holdgate, M. 2003. *Penguins & Mandarins. The Memoirs of Martin Holdgate*. The Memoir Club, Whitworth Hall, Spennymoor, Co. Durham.

Holman, J. 1835. *A Voyage round the World including Travels in Africa, Asia, Australia, America (etc. etc.) from 1827 to 1832*. Vol. 3. Smith, Elder & Co, London.

Im Thurn, E. & L. Wharton, 1925. *The Journal of the William Lockerby, Sandalwood Trader in the Fijian Islands during the Years 1808-09*. Hakluyt Society, London.

Landwehr, J. 1991. *VOC, a Bibliography of Publications relating to the Dutch East India Company 1602-1800*. HES, Utrecht.

Levell, S. *Cycling Across the South Atlantic. An Oceanic Odyssey with a Bicycle*. George Mann Publications, Easton, Winchester.

Linklater, E. 1972. *The Voyage of the Challenger*. John Murray, London.

Lopez de Castanheda, F. 1833. *História do descobrimento e conquista da India pelos portugueses*. Typographia Rollandia, Lisboa.

Mainwaring, M. 1988. *Nor Any Drop to drink. England to Australia May 1987-January 1988*. Bloomsbury Publishing Ltd., London.

Major, R.H. 1857. *Early Voyages to Terra Australis at the Time of Captain Cook as told in Original Documents*. Hakluyt Society, London. Reprint Australian Heritage Press, Adelaide 1963, Franklin, New York 1970.

Malhão Pereira, J.M. 2001. *Um Livro de Marinharia do século XVIII - Estudo Critico*. Thesis Lisbon University.

Marr, J.W.S. 1923. *Into the Frozen South*. Cassell & Co., London.

McPherson, S. 2016. *Britain's Treasure Islands. A Journey to the UK Overseas Territories*. Redfern Natural History Productions, Poole.

Meilink-Roelofs, M.A.P., R.Raben & H. Spijkerman (ed.) 1992. *The Archives of the Dutch East India Company (1602-1795)*. SDU-Uitgeverij, The Hague.

Milner, J. & O.W. Brierly, 1869. *The Cruise of H.M.S. 'Galatea', Captain H.R.H. the Duke of Edinburgh, K.G. in 1867-1868*. W.H. Allen, London.

Moreland, W.H. 1934. *Peter Floris, his Voyage to the East Indies in the Globe 1611-1615*. Hakluyt Society, London.

Morrell, A.J. 1833. *Narrative of a Voyage to the Ethiopic and South Atlantic Ocean, Indian Ocean, Chinese Sea, North and South Pacific Ocean, in the years 1829, 1830, 1931*. Harper, New York. Reprint by Gregg Press, Upper Saddle River, 1970.

Morrell, B. 1832. *A Narrative of Four Voyages, to the South Sea, North and South Pacific Ocean, Chinese Sea, Ethiopic and Southern Atlantic Ocean, Indian and Antarctic Ocean. From the Year 1822 to 1831*. Harper, New York. Reprint by Gregg Press, Upper Saddle River, 1970.

Moseley, H.N. 1879. *Notes by a Naturalist, being an Account of Various Observations made during the Voyage of H.M.S. 'Challenger', round the World in the Years 1872-1876*. Methuen, London. Reprint by Laurie, London, 1944.

Newby, E. 1956. *The Last Grain Race*. Secker & Warburg.

Nicoll, M.J. 1908. *Three Voyages of a Naturalist; being an Account of Many Little-Known Islands in the Three Oceans visited by the 'Valhalla' R.Y.S.* Witherby & Co., London.

Oldfield, S. 1987. *Fragments of Paradise: a Guide for Conservation Action in the UK Dependent Territories.* Pisces for British Association of Nature Conservationists, Oxford.

Oliver, P. 1891. The *Voyage of François Leguat to Rodriguez*, vol. 1. Hakluyt Society, London.

Perrone-Moisés, L. 1992. *Vinte Luas. Viagem de Paulmier de Gonneville ao Brasil: 1503-1505.* Schwarcz, São Paulo.

Purdy, J. 1816. *The Oriental Navigator.* James Whittle & Richard Holmes Laurie, London. Online: https://books.google.nl/books?id=aL2YtgEACAAJ&printsec=frontcover&hl=nl&source=gbs_ge_summary_r&cad=0#v=onepage&q&f=false

Ridgeway, R., M.C. Ridgeway & R. Ridgeway, 1996. *Then we sailed away.* Little, Brown and Company, London.

Ritchie, H. 1997. *The Last Pink Bits. Travels Through the Remnants of the British Empire.* Hodder & Stoughton, London.

Roach, P. 1952. *Voyage in a Barquentine.* Rupert Hart-Davis, London.

Schaffer, I. 2010. *The Sea Shall Not Have Them. Narrative of Stephen and Margaret White, who were shipwrecked near Tristan da Cunha on the Blenden Hall in 1821, and their arrival in Van Diemen's Land 1832.* Private publication, Tasmania.

Schilder, G. (ed.) 1976. *Voyage to the Great South Land, Willem de Vlamingh 1696-1697.* Royal Australian Historical Society, Sydney.

Seligman, A. 1939. *The Voyage of the 'Cap Pilar'*. Hodder & stoughton, London. Reprint by Seafarer Books, London, 1993.

Soeiro de Brito, J., M.N. Teague & J. Brandão (co-ordinators) 1992. *O Livro de Lisuarte de Abreu*. Comissão Nacional para as Comemorações dos Descobrimentos Portugueses, Lisboa.

Spry, W.J.J. 1877. *The Cruise of Her Majesty's Ship 'Challenger'*. Samson, Low, Maston, Searle & Rivenzta, London.

Stapel, F.W. 1927. *Pieter van Dam's Beschryvinge van de Oostindische Compagnie*. Martinus Nijhoff, The Hague.

Staunton, G.T. 1797. *An Authentic Account of an Embassy from the King of Great Britain to the Emperor of China taken chiefly from the Papers of His Excellency the Earl of Macartney*. Vol. 1. Nicol, London.

Stevenson, R.E. & F.H. Talbot (eds.) 1994. *Islands. The Illustrated Library of the Earth*. Time-Life Books, Amsterdam.

Stirling, W. 1843. *Narrative of the Wreck of the Ship 'Tiger', of Liverpool (Capt. Edward Searight), on the Desert Island of Astova, on the Morning of the 12th of August 1836*. Roberts, Exeter.

Swire, H. 1938. *The Voyage of the 'Challenger': a Personal Narrative of the Historic Circumnavigation of the Globe in the Years 1872-1876*. Golden Cockerel Press, London.

Taylor, G. 1978. *The Sea Chaplains*. Oxford Illustrated Press, Oxford.

Thomson, C.W. 1875. *The Voyage of the Challenger*. Vol. 1, *Narrative and Part I*. Macmillan, London.

Thomson, C.W. 1877. *The Voyage of the Challenger*. Vol. 2, *The Atlantic*. Macmillan, London.

Thrower, N.J.W. 1980. *The Three Voyages of Edmund Halley in the Paramore 1698-1701*. Hakluyt Society, London.

Tiltman, M.H. 1933. *God's Adventurers*. George G. Harrap & Co Ltd, London.

Uhden, R. 1939. The Oldest Portuguese Original Chart of the Indian Ocean, A.D. 1509. *Imago Mundi* **3**: 7-11.

Villers, A. 1940. *Cruise of the Conrad*. Hodder & Stoughton, London.

Watts, C.C. 1936. *In Mid-Atlantic: the Islands of St Helena, Ascension and Tristan da Cunha*. Society for the Propagation of the Gospel, London.

Wild, F. 1923. *Shackleton's Last Voyage. The Story of the 'Quest'*. Cassell, London.

Wild, J.J. 1878. *At Anchor. A Narrative of Experiences Afloat and Ashore during the Voyage of H.M.S. 'Challenger' from 1872 to 1876*. Marcus Ward, London.

Williamson, C. 1961. *Great True Stories of the Islands*. Arco Publications, London.

Winchester, S. 1985. *Outposts. Journeys to the Surviving Relics of the British Empire*. Hodder & Stoughton, London.

Wytema, M.S. 1936. *'Klaar voor onder water!' Met Hr. Ms. 'K-XVIII' langs een omweg naar Soerabaia*. Andries Blitz, Amsterdam.

3. **Fiction**

Bazin, H. 1970. *Les Bienheureux de la désolation*. Editions du Seuil, Paris. English translation: *Tristan, a novel*. Simon & Shuster, London, 1971.

Campbell, R. 1949. 'Tristan da Cunha'. In: *The Collected Poems of Roy Campbell*. The Bodley Head, London.

Harris, Z. 2000. *Further than the Furthest Thing*. Faber & Faber Ltd., London.

Holtby, W. 1933. *The Astonishing Island. Being a Veracious Record of the Experiences Undergone by Robinson Lippingtree Mackintosh from Tristan da Cunha during an eccidental Visit to Unknown Territory in the Year of Grace MCMXXX-?* Macmillan Company, New York.

Jenkins, G. 1962. *A Grue of Ice*. Collins, London.

Levi, P. 1975. *Il Systema periodico*. Giulio Einaudi editore S.p.A., Torino. English translation: *The Periodic Table*, Schocken Books, 1984.

Marsh, J.E. 1950. *On the Trail of the Albatross*. Burke Publishing, Christchurch UK.

Newman, G. 1900. *Other Lyrics*. Reprint by Forgotten Books, London, 2018.

Schrott, R. 2003. *Tristan da Cunha, oder Die Hälfte der Erde*. Carl Hanser Verlag, München.

Townsend, J.R. 1981. *The Islanders*. Oxford University Press, Oxford.

Townsend, J.R. 1992. *The Invaders*. Oxford University Press, Oxford.

Verne, J. 1867-1868. *Les enfants du capitaine Grant*. Hetzel, Paris. English translation: *A Voyage Round The World*. Routledge & Sons 1876. Many later reprints: *In search of the Castaways; Or, The Children of Captain*

Grant. Online available at
http://forgottenfutures.com/library/grant/grant.htm
(book 2, chapter 2 on Tristan).

Watt, D. 1991. *Trouble in Tristan*. Tynron Prss,
Stenhouse, Thornhill, Scotland.

4. **Moorhen sources**

Apart from the papers listed below, Moorhens have been
mentioned by Early (1833, so also in McCormick, 1966),
Frazet *et al*. (1983), Hagen (1952), Im Thurn & Wharton
(1925), Milner & Brierly (1869), Morrell (1832), Moseley
(1879), Purdy (1816), Stirling (1843), Swales *et al*. (1993),
Thomson (1877) and Wace & Holdgate (1976). And of
course by myself (Beintema 1997).

Allen, J.A. 1892. Description of a new Gallinule from
Gough Island *Bull. Am. Mus. Nat. Hist*. **4**: 57-58.

Beintema, A.J. 1972. The History of the Island Hen
(*Gallinula nesiotis*), the extinct Flightless Gallinule of
Tristan da Cunha. *Bull. B.O.C*. **92**: 106-113.

Bourne, W.R.P. & A.C.F. David, 1981. Nineteenth
Century Bird Records from Tristan da Cunha. *Bull.
B.O.C*. **101**: 247-256.

Brierly, O. 1842. *Diary kept while Aboard R.Y.S.
Wanderer (Capt. Bushby, R.N.), voyaging from
Plymouth to Australia*. manuscript in Mitchell Library,
Sidney.

Broekhuysen, G.J. & W. Macnae, 1949. Observations on
the Birds of Tristan da Cunha Islands and Gough Island
in February and early March 1948. *Ardea* **87**: 97-113.

Brooke, R.K. 1979. Some mid-XIX Century Bird
Collections from Tristan da Cunha. *Cormorant* **7**: 24-
26.

Bullock, W. 1819. *Catalogue (without which no person can be admitted either to the view or the sale) ofthe Roman Gallery of Antiquities and Works of Art, and the London Museum of Natural History* (etc. etc.). Annotated copy in Museum of Zoology, Cambridge.

Busk, G. 1869. (On additions to the menagerie). *Proc. Zool. Soc. London* 1869: 469.

Carmichael, D. 1818. Some Account of the Island of Tristan da Cunha and of its Natural Productions. *Transactions Linnean Soc*. **12**: 483-513.

Eber, G. 1961. Vergleichende Untersuchungen am flugfähigen Teichhhuhn *Gallinula chl. chloropus* und der flugunfähigen Inselralle *Gallinula nesiotis*. *Bonner Zool. Beitr*. **12**: 247-315.

Elliot, H.F.I. 1957. A contribution to the ornithology of the Tristan da Cunha Group. *Ibis* **99**: 545-586.

Gadow, H. 1910. On the ornithological Collections of the University of Cambridge. *Ibis* (9) **4**: 47-53.

Groenenberg D.S.J, A.J. Beintema, R.W.R.J. Dekker & E. Gittenberger, 2008. Ancient DNA Elucidates the Controversy about the Flightless Island Hens (*Gallinula* sp.) of Tristan da Cunha. *PLoS ONE* 3(3): e1835. Online: https://doi.org/10.1371/journal.pone.0001835.

Gurney, J.H. 1853. (On a wingless Bird of Tristan da Cunha). *Zoologist*: 4017.

Layard, E.L. 1856. (Note on specimens collected by Captain Nolloth). *Nautical Mag*. **25**: 414-415.

Layard, E.L. 1869. Further Notes on South African Ornithology. *Ibis* (2) **5**: 361-378.

Macgillivray, J. 1852. *Journal kept Aboard H.M.S. Herald, November 11th-13th 1852*. Manuscript in Admiralty Library, London.

Matthews, G.M. 1932. The Birds of Tristan da Cunha. *Novitates Zo*ol. **38**: 13-48.

Nicoll, M.J. 1906. On the Birds collected and observed during the Voyage of the 'Valhalla'. *Ibis* (8) **6**: 666-712.

Nolloth, M.S. 1856. Visit of H.M.S. 'Frolic' to Tristan da Cunha. *Nautical Mag.* **25**: 400-413.

Olson S. 1973. Evolution of the Rails of the South Atlantic Islands (Aves: Rallidae). *Smithsonian Contrib. Zool.* **152**: 1-53.

Peringuey, L. 1910. (On collections from Tristan da Cunha). *Report of the South African Museum for the year ended 31st December 1909*: 7-8.

Richardson, M.E. 1984. Aspects of the Ornithology of the Tristan da Cunha Group and Gough Island, 1972-1974. *Cormorant* **12**: 123-201.

Sclater, P.L. 1861. On the Island-Hen of Tristan d'Acunha. *Proc. Zool. Soc. London* 1861: 260-263.

Sperling, R.M. 1872. Letter on Tristan da Cunha. *Ibis* (3) **2**: 74-79.

Stresemann, E. 1953. Birds collected by Capt. Dugald Carmichael on Tristan da Cunha 1816-1817. *Ibis* **95**: 146-147.

Warren, R.L.M. 1966. *Type Specimens of Birds in the British Museum (Natural History)*. Vol 1. Non-Passerines. British Museum (Nat. Hist.), London.

Watkins, B.P. & R.W. Furness, 1986. Population Status, Breeding and Conservation of the Gough Moorhen. *Ostrich* **57**: 32-36.

Willemoes Suhm, R. von, 1876. On Observations made during the Voyage of H.M.S. 'Challenger'. *Proc. Royal Soc.* **24**: 569-585.

Wilson, A.E. & M.K. Swales, 1958. Flightless Moorhens (*Porphyriornis c. comeri*) from Gough Island breed in Captivity. *Avicultural Mag.* **64**: 43-45.

Winterbottom, J.M. 1958. Tristan da Cunha Birds. *Ibis* **100**: 285.

Winterbottom, J.M. 1976. Keytel's Birds from Tristan da Cunha. *Ostrich* **47**: 69-70.

Lightning Source UK Ltd.
Milton Keynes UK
UKHW052006050922
408368UK00001B/30

9 781803 692555